Decoherence Suppression
in Quantum Systems
2008

Kinki University Series on Quantum Computing

Editor-in-Chief: Mikio Nakahara *(Kinki University, Japan)*

ISSN: 1793-7299

Published

Vol. 1   Mathematical Aspects of Quantum Computing 2007
edited by Mikio Nakahara, Robabeh Rahimi (Kinki Univ., Japan) &
Akira SaiToh (Osaka Univ., Japan)

Vol. 2   Molecular Realizations of Quantum Computing 2007
edited by Mikio Nakahara, Yukihiro Ota, Robabeh Rahimi, Yasushi Kondo &
Masahito Tada-Umezaki (Kinki Univ., Japan)

Vol. 3   Decoherence Suppression in Quantum Systems 2008
edited by Mikio Nakahara, Robabeh Rahimi & Akira SaiToh
(Kinki Univ., Japan)

Kinki University Series on Quantum Computing – Vol. 3

editors

Mikio Nakahara
Robabeh Rahimi
Akira SaiToh

Kinki University, Japan

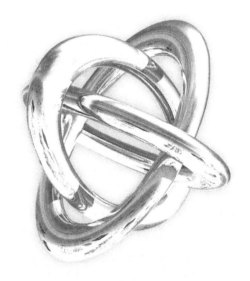

Decoherence Suppression
in Quantum Systems
2008

World Scientific

NEW JERSEY · LONDON · SINGAPORE · BEIJING · SHANGHAI · HONG KONG · TAIPEI · CHENNAI

Published by

World Scientific Publishing Co. Pte. Ltd.
5 Toh Tuck Link, Singapore 596224
USA office: 27 Warren Street, Suite 401-402, Hackensack, NJ 07601
UK office: 57 Shelton Street, Covent Garden, London WC2H 9HE

British Library Cataloguing-in-Publication Data
A catalogue record for this book is available from the British Library.

DECOHERENCE SUPPRESSION IN QUANTUM SYSTEMS 2008
Proceedings of the Symposium

ISBN-13 978-981-4295-83-3
ISBN-10 981-4295-83-3

Printed in Singapore.

PREFACE

This volume contains lecture notes and a paper contributed from speakers of the Symposium on Decoherence Suppression in Quantum Systems held at Oxford Kobe Institute, Kobe, Japan, from 7th to 11th September 2008. The symposium was aimed to gather researchers working or having interests in the field of decoherence and its suppression in order to evoke further developments in the field and to extend mutual collaborations.

The symposium was supported by "Open Research Center" Project for Private Universities: matching fund subsidy from MEXT (Ministry of Education, Culture, Sports, Science and Technology).

This volume is organized so that a reader can learn many aspects of decoherence and its suppression methods from fundamentals to recent topics by following the lecture notes one by one. It is supposed to be useful for a wide range of audience from a graduate student to a researcher with interest in this field, as the lectures note have been prepared in a self-contained manner.

It is our hope that continuous and active researches in this field will eventually resolve the long-standing problems of decoherence in order for early realization of practical quantum computing.

We would like to thank Ms Zhang Ji of World Scientific for her editorial support.

Higashi-Osaka, September 2009

Mikio Nakahara
Robabeh Rahimi
Akira SaiToh

OUTLOOK OF THE SYMPOSIUM

Title	Symposium on Decoherence Suppression in Quantum Systems
Place	Oxford Kobe Institute, Kobe, Japan
Schedule	7-10 September 2008
Committee	Mikio Nakahara, Kinki University, Japan Robabeh Rahimi, Kinki University, Japan Yukihiro Ota, Kinki University, Japan Akira SaiToh, Kinki University, Japan
Invited Speakers	Gen Kimura, National Institute of Advanced Industrial Science and Technology, Japan Frank Gaitan, Southern Illinois University, USA Yasushi Kondo, Kinki University, Japan Yukihiro Ota, Kinki University, Japan Akira SaiToh, Kinki University, Japan Osama Moussa, Waterloo University, Canada
Contributed Speaker	Masamitsu Bando, Osaka University, Japan
URL	http://alice.math.kindai.ac.jp/~symp08/

PROGRAM

7 (Sun) September 2008

10:45-12:15	Reception
12:15-13:00	Lunch
14:00-15:30	G. Kimura, *Introduction to Decoherence: Overview* I
15:30-16:00	Coffee Break
16:00-17:30	G. Kimura, *Introduction to Decoherence: Overview* II
18:00-19:00	Supper
20:00-21:30	G. Kimura, *Introduction to Decoherence: Overview* III

8 (Mon) September 2008

7:45-8:45	Breakfast
9:00-10:30	F. Gaitan, *Quantum Error Correction and Fault-Tolerant Quantum Computing* I
10:30-10:45	Coffee Break
10:45-12:15	F. Gaitan, *Quantum Error Correction and Fault-Tolerant Quantum Computing* II
12:15-13:00	Lunch
14:00-15:30	F. Gaitan, *Quantum Error Correction and Fault-Tolerant Quantum Computing* III
15:30-16:00	Coffee Break
16:00-17:30	Y. Kondo, *Engineered Noise: Application to Decoherence Suppression Experiments*
18:00-19:00	Supper
20:00-21:30	Y. Ota, *Composite Pulses as Geometric Quantum Gates*

9 (Tue) September 2008

7:45-8:45	Breakfast
9:30-10:30	O. Moussa, *Noise Characterization and Quantum Error Correction in Solid-state NMR* I
10:30-10:45	Coffee Break
11:00-12:00	O. Moussa, *Noise Characterization and Quantum Error Correction in Solid-state NMR* II
12:15-13:00	Lunch
14:00-15:30	F. Gaitan, *Quantum Error Correction and Fault-Tolerant Quantum Computing* IV
15:30-16:00	Coffee Break
16:00-16:20	M. Bando, *Holonomic quantum computing in a dimer model*
18:00-19:00	Supper
20:00-21:30	Informal Discussions

10 (Wed) September 2008

7:45-8:45	Breakfast
9:00-10:30	A. SaiToh, *Quantum Wipe Effect* I
10:30-10:45	Coffee Break
10:45-12:15	A. SaiToh, *Quantum Wipe Effect* II
12:15-13:00	Lunch
14:00-	Ending

LIST OF PARTICIPANTS

Daisuke Aratsu	Osaka University, Japan
Chiara Bagnasco	Università di Roma, La Sapienza, Italy
Masamitsu Bando	Osaka University, Japan
Vahideh Ebrahimi Bakhtavar	Kinki University, Japan
Frank Gaitan	Southern Illinois University, USA
Kenichi Kasamatsu	Kinki University, Japan
Gen Kimura	National Institute of Advanced Industrial Science and Technology, Japan
Yasushi Kondo	Kinki University, Japan
Osama Moussa	University of Waterloo, Canada
Ayumi Muramatsu	Kinki University, Japan
Mikio Nakahara	Kinki University, Japan
Tetsuo Ohmi	Kinki University, Japan
Yukihiro Ota	Kinki University, Japan
Robabeh Rahimi Darabad	Kinki University, Japan
Akira SaiToh	Kinki University, Japan
Masahito Tada	Kinki University, Japan
Masakazu Tanaka	Waseda University, Japan
Hiroyuki Tomita	Kinki University, Japan

CONTENTS

ELEMENTARY MATHEMATICAL FRAMEWORK FOR OPEN QUANTUM *d*-LEVEL SYSTEMS: DECOHERENCE OVERVIEW

GEN KIMURA *

Research Center for Information Security (RCIS), National Institute of Advanced Industrial Science and Technology (AIST), Daibiru building 1102, Sotokanda, Chiyoda-ku, Tokyo, 101-0021, Japan.
* E-mail: gen-kimura[at]aist.go.jp

This lecture note provides an easy introduction of the theory of open quantum systems in a physically and mathematically closed manner. After a compact review of quantum mechanics, we explain how to treat open quantum systems which turns out to explain the decoherence process. In order to be logically closed, we restrict to the finite quantum systems, but almost all the mathematical techniques are explained in detail so that the students can follow the equations and master the techniques which are usually assumed from the beginning. In particular, there are almost 90 exercises which supplement the contents to understand, and all the solutions will be uploaded in my web page*.

Keywords: Open Quantum Systems; Completely Positive Dynamical Semigroup

1. Decoherence in open quantum systems

It has passed almost one century after the discovery of quantum theory, and people has gradually noticed a powerful application of a coherent usage of quantum mechanics, especially for several information processings. These includes quantum computation, quantum cryptography, and quantum teleportation, and they are expected to be near-future technologies.[1] However, quantum mechanics is basically the theory of the microscopic world and is usually quite fragile against the noise. Therefore, it is usually difficult to realize quantum devices in order to use them for information processings as are proposed in an ideal situation. Many attempts are in progress to overcome the undesirable noise, namely the process of which is usually called

---

*You can find this at http://staff.aist.go.jp/gen-kimura/decoherence.pdf.

a **decoherence**,[2] so that the designed devices work in coherence with the theoretical prediction of the quantum mechanics in the ideal situation.

Historically, the original motivation of the studies on decoherence was to explain how classical world appears from quantum world.[3,4] (See Refs. 2 and 5 for the overview of decoherence from this point of view.) In the typical scales in our daily life, it seems that no quantum effects are effective, and the classical theory is sufficient to explain the phenomena around us. On the other hand, it is undeniable that all the material are composed of atoms, the behaviors of which are governed by the law of quantum mechanics. Therefore, it is for sure one of the important subjects to understand where and how the classical nature appears from quantum world. In the meanwhile, the interpretation of quantum theory is still a controversy problem and this sometimes makes the problem of decoherence not a physical problem but a philosophical problem, which might root in the belief of each physicist. (See, for instance, Ref. 6 for the problem of decoherence in connection with several interpretation of quantum mechanics.) Although it is quite interesting to challenge in one's own way for every physicists, but there pervades a (bad) convention in physics community to persuade young researchers not to proceed into this direction. Nevertheless, it is still possible to generally formulate the problem of decoherence in a formal manner as a study of the discord from the ideal unitary dynamics (for an isolated quantum system), while mostly, the reason of the discord is explained by the interaction with the environment. This attitude is not bad at least from the practical and operational point of view. In particular, for people who are willing to realize the quantum device for such as quantum computations, the main obstacle is a quantum noise from the undesirable interaction with the environment, and thus nowadays people sometimes identify the notion of decoherence and a quantum noise.[1] Therefore, the study of decoherence is closely related and can be almost identified with the theory of open quantum systems,[8-14] where we take account of the effects of the environment to explain the phenomena in the system of interest.

In this note, we introduce the theory of open quantum systems both from physical and mathematical point of view. Our purpose is to provide a mathematical framework in order to deal with decoherence processes in logically closed manner as much as possible. Indeed, the study of open system sometimes involves essentially infinity (infinite dimension of Hilbert space, infinite degrees of freedom) to explain a time irreversible process of the system. Therefore, to deal with the theory fully with mathematical rigor sometimes keeps physicists and young students off from the theory.

That is not our intention and therefore we restrict to finite dimensional quantum systems (a d-level system), and will explain necessary notions for studying open quantum systems in both physically and mathematically closed way. The readers can refer the appendix A whenever they come to unfamiliar mathematical notions. In additions, there are almost 90 exercises to supplement the contents. We suggest that students should try to solve all the exercises by themselves except for a few difficult ones, but all the solutions will be uploaded in the web page <http://staff.aist.go.jp/gen-kimura/decoherence.pdf> for the readers' convenience. We believe that the restriction to the finite cases will not dramatically lose the essence of the theory of open quantum systems but by contrast it will provide an useful introduction of the theory where one does not bother over irrelevant difficulties of mathematics to deal with infinity and complex topological problems.

This note is organized as follows. In Sec. 2, we start from the detailed review of the quantum theory for d-level systems. In particular, the structure of the state space will be investigated. In Sec. 3, a phenomenological consideration of quantum dynamics will be addressed and a general time evolution map is shown to be a trace preserving completely positive map. In Sec. 4, the theory of open quantum systems is explained in detail by combining the phenomenological and microscopic points of view. The general class of entropy non-decreasing process is discussed. Finally in order for the general treatment of quantum Markovian process, we explain the representation theorem (Gorini-Kossakowski-Lindblad-Sudarshan master equation) for a completely positive dynamical semigroup.

2. Quantum Mechanics for d-level systems

In this section, we provide a compact explanation of the theory of quantum mechanics, restricting to the finite level systems. In almost any physical theory, including quantum theory, the notions of system (including composition of several systems), states, observables, and dynamics play an important role as basic ingredients of the theory. In this section, these notions in quantum mechanics in finite level systems are explained in order, raising the basic postulates P1 — P5 of quantum mechanics. The readers should refer the mathematical background in Appendix A whenever they encounter the unfamiliar mathematical notions.

[P1: Quantum System] *For each quantum system, there exists the associated Hilbert space* $\mathcal{H}$ *on which all the phenomena for the system are* *theoretically described.*

In this note, we deal with only d-**level quantum system** where the associated Hilbert space — complete inner product — has a finite dimension $d \in \mathbb{N}$. In the field of quantum information theory, a 2-level system is called a **qubit** system, while a general d-level system is often called a **qudit** system. Hereafter, $\mathcal{H}$ denotes a d-dimensional Hilbert space (inner product space) for some $d \in \mathbb{R}$ with an inner product $\langle \psi, \phi \rangle \in \mathbb{C}$ between $\psi, \phi \in \mathcal{H}$ [*1].

[Example] d level system is (essentially) described by $\mathbb{C}^d$. The inner product between $\boldsymbol{a} = (a_1, \ldots, a_d)^T$, $\boldsymbol{b} = (b_1, \ldots, b_d)^T \in \mathbb{C}^d$ is given by $\langle \boldsymbol{a}, \boldsymbol{b} \rangle = \sum_{i=1}^{d} \overline{a_i} b_i$ (see Exe. 54 and also Exe. 58).

[P2: Quantum State] *Quantum state is represented by a density operator.*

A **density operator** is a positive linear operator[*2] $\rho \geq 0$ on $\mathcal{H}$ with a unit trace $\operatorname{tr} \rho = 1$. Here, we implicitly assume that all density operators can be prepared in principle, and we shall identify a state and the corresponding density operator.

Exercise 1. Show that the eigenvalues p_i $(i = 1, \ldots, d)$ of any density operator is a probability distribution: i.e., $p_i \geq 0$ $(i = 1, \ldots, d)$, $\sum_{i=1}^{d} p_i = 1$.

For a unit vector $\psi \in \mathcal{H}$, an operator of the form $\rho_\psi := |\psi\rangle\langle\psi|$ — i.e., one dimensional projection operator onto the subspace spanned by ψ — is a density operator[*3]. A density operator ρ_ψ of this form is called a **pure state**, while the unit vector ψ is called a **state vector**. A density operator which can not be written in this form is called a **mixed state**. As we shall see below, pure states play a fundamental role in quantum state. We occasionally identify a state vector $\psi \in \mathcal{H}$ and the corresponding density operator $\rho_\psi = |\psi\rangle\langle\psi|$. Since state vectors $\psi, \phi \in \mathcal{H}$ can be mathematically

---

[*1] Notice the different convention for the inner product in physics and mathematics: In physics, the inner product is usually linear in the second argument, and is antilinear in the first argument: $\langle \psi, \alpha\psi + \beta\xi \rangle = \alpha\langle \psi, \phi \rangle + \beta\langle \psi, \xi \rangle$ for any $\psi, \phi, \xi \in \mathcal{H}$, $\alpha, \beta \in \mathbb{C}$. In mathematics, the convention is usually the opposite. In this note, we follow the physics convention since it is symbolically clever when dealing with the operators of the form $|\psi\rangle\langle\phi|$ (see Appendix A for this notation.) to visually keep the "associative law": $|\psi\rangle\langle\phi|\xi = \psi\langle\phi, \xi\rangle$.

[*2] A linear operator ρ on $\mathcal{H}$ is called positive (or positive semidefinite), symbolically denoted as $\rho \geq 0$, iff $\langle \psi, \rho\psi \rangle \geq 0$ for any $\psi \in \mathcal{H}$. It turns out that ρ is positive if and only if ρ is an Hermitian operator with positive eigenvalues (see Exercise 67).

[*3] Indeed, for any vector $\phi \in \mathcal{H}$, it follows $\langle \phi, \rho_\psi \phi \rangle = |\langle \phi, \psi \rangle|^2 \geq 0$, and $\operatorname{tr} \rho_\psi = \langle \psi, \psi \rangle = \|\psi\|^2 = 1$ (see Exe. 82).

added as vectors, there exists the corresponding state vector $\psi + \phi$ (with a suitable normalization). This is called a **superposition principle** in quantum mechanics, which is one of the most remarkable features of quantum mechanics. One should notice that there exists a superposed state[*4] even for two states which are mutually exclusive states, e.g., states with switch on and off.

[Example] In qubit system $\mathcal{H} \simeq \mathbb{C}^2$, let $\psi_1 = (1,0)^T$, $\psi_2 = (0,1)^T$, $\psi_3 = \frac{1}{\sqrt{2}}(1,1)^T$, and $\psi_4 = \frac{1}{\sqrt{2}}(1,-1)^T$. Since all of these are unit vectors, they can be considered as state vectors, with the corresponding density operators $\rho_{\psi_i} := |\psi_i\rangle\langle\psi_i|$:

$$\rho_{\psi_1} = \begin{pmatrix} 1 & 0 \\ 0 & 0 \end{pmatrix}, \quad \rho_{\psi_2} = \begin{pmatrix} 0 & 0 \\ 0 & 1 \end{pmatrix}, \quad \rho_{\psi_3} = \frac{1}{2}\begin{pmatrix} 1 & 1 \\ 1 & 1 \end{pmatrix}, \quad \rho_{\psi_4} = \frac{1}{2}\begin{pmatrix} 1 & -1 \\ -1 & 1 \end{pmatrix}.$$

In general, a unit vector $\psi = (c_1, \ldots, c_d)^T \in \mathbb{C}^d$ ($\sum_i |c_i|^2 = 1$) for a qudit system has the corresponding density operator:

$$\rho_\psi = (\overline{c_1}, \ldots, \overline{c_d}) \begin{pmatrix} c_1 \\ \vdots \\ c_d \end{pmatrix} = \begin{pmatrix} \overline{c_1}c_1 & \cdots & \overline{c_d}c_1 \\ \vdots & \ddots & \vdots \\ \overline{c_1}c_d & \cdots & \overline{c_d}c_d \end{pmatrix}.$$

Examples for mixed states in qubit system are

$$\rho_1 = \frac{1}{2}\begin{pmatrix} 1 & 0 \\ 0 & 1 \end{pmatrix} = \frac{1}{2}\mathbb{I}, \quad \rho_2 = \frac{1}{4}\begin{pmatrix} 2 & -1 \\ -1 & 2 \end{pmatrix}, \quad etc. \tag{1}$$

One should check that they are indeed density operators by showing the Hermicity with positive eigenvalues — i.e., they are positive — with unit trace.

[P3: Observables] *Physical quantity is represented by an Hermitian operator.*

In quantum mechanics, physical quantities are called **observables**.[15] In this note, we assume that any Hermitian operator has the corresponding observable in principle, and we shall identify an observable and the corresponding Hermitian operator.

So far, we have not yet described any physical contents available for experiments, but just have introduced the (mathematical) tools to represent states and observables. Together with states and observables, we obtain the physical information from these mathematical objects as follows:

[*4]There are many attempts to interpret such superposed states, but even now there is no agreement in opinions among physicists in the interpretation of superposition principle.

[Born's statistical formula] *If one measures an observable O under a state ρ, then one observes an outcome $\lambda_i \in \mathbb{R}$, which is one of the eigenvalues of O, with the probability $\Pr\{\lambda_i \parallel O, \rho\}$ given by*

$$\Pr\{\lambda_i \parallel O, \rho\} = \mathrm{tr}[P_i \rho]. \qquad (2)$$

Here P_i denotes the spectral projection of O corresponding to eigenvalue λ_i.

Remind here the spectral decomposition $O = \sum_i \lambda_i P_i$ (see Appendix A). Notice that all eigenvalues of any Hermitian operator are real (see Exercise 64). This reflects that experimental data are generally real.

Exercise 2. Prove that $\mathrm{tr}[P_i \rho]$ gives a probability distribution, i.e., $\mathrm{tr}[P_i \rho] \geq 0$, $\sum_i \mathrm{tr}[P_i \rho] = 1$.

One can say that quantum mechanics is essentially the theory of probability[*5], where the probability is described by Born's statistical formula with the fixed context of state and observable. The reproducibility condition is usually assumed so that any experiment in principle should be repeated and reproducible. In other words, we implicitly assume that independent and identically distributed (i.i.d.) states can be prepared in principle, and by observing the same observable O for each state ρ, experimentalists can observe the probability distribution, and confirm the validity of Born's formula.

We say observable O is **nondegenerate** if the multiplicities of all eigenvalues are 1, i.e., the dimensions of all the eigenspaces equal 1; otherwise, O is called **degenerate**. Although Born's statistical formula (2) is the most general, one should recognize each formula corresponding to each combination of pure states and nondegenerate observables as follows:

- [For pure state $\rho_\psi = |\psi\rangle\langle\psi|$ and general observable $O = \sum_i \lambda_i P_i$]

$$\Pr\{\lambda_i \parallel O, \rho\} = \langle \psi, P_i \psi \rangle. \qquad (3)$$

- [For general state ρ and nondegenerate observable $O = \sum_{i=1}^d \lambda_i |\psi_i\rangle\langle\psi_i|$]

$$\Pr\{\lambda_i \parallel O, \rho\} = \langle \psi_i, \rho \psi_i \rangle, \qquad (4)$$

[*5]It seems not reasonable to seek the *hidden* origin of the probability of quantum mechanics if we require some appropriate conditions. Readers who are interested in this direction should read, for instance, Refs. 16,17, for EPR paradox, Kochen-Specker theorem, and Bell's theorem, etc.

where ψ_i is a normalized eigenvector of O corresponding to the eingenvalue λ_i.

- [For pure state $\rho_\psi = |\psi\rangle\langle\psi|$ and nondegenerate observable $O = \sum_{i=1}^d \lambda_i|\psi_i\rangle\langle\psi_i|$]

$$\Pr\{\lambda_i \,\|\, O, \rho_\psi\} = |\langle\psi_i, \psi\rangle|^2. \tag{5}$$

The final case might be the prototype of quantum mechanics which often appears in the beginning of the standard textbook of quantum mechanics. Readers are recommended to remember each formula (2), (3), (4), and especially (5), and use the most convenient one in each context.

Exercise 3. Show (3), (4), and (5) from (2).

Remind that the set of eigenvectors $\{\psi_i\}_{i=1}^d$ of observable O forms an orthonormal basis of $\mathcal{H}$, since any Hermitian operator can be diagonalized by a unitary operator. Therefore, any state vector $\psi \in \mathcal{H}$ can be written as

$$\psi = \sum_{i=1}^d \langle\psi_i, \psi\rangle\psi_i. \tag{6}$$

By comparing (6) with (5), one should observe that the absolute square of each coefficient $\langle\psi_i, \psi\rangle$ in the expansion (6) of a state vector gives a probability (5) to get the corresponding outcome λ_i. In this sense, $\langle\psi_i, \psi\rangle$ is sometimes called a **probability amplitude**.

Since we are dealing with probability, familiar notions of expectation value (mean value), variance, standard deviation (root-mean-square deviation), etc. in probability theory are also defined in Quantum mechanics as well: Let us denote expectation values, variance, and standard deviation of observable O under state ρ by $E[O]_\rho$, $V[O]_\rho$, and $\sigma[O]_\rho := \sqrt{V[O]_\rho}$, respectively, which are defined by

$$E[O]_\rho := \sum_i \lambda_i \Pr\{\lambda_i\|O, \rho\}, \tag{7}$$

$$V[O]_\rho := \sum_i (\lambda_i - E[O]_\rho)^2 \Pr\{\lambda_i\|O, \rho\}. \tag{8}$$

From the spectral decomposition $O = \sum_i \lambda_i P_i$ and Born's statistical formula (2), we obtain a simple formula for an expectation value:

$$E[O]_\rho = \mathrm{tr}[O\rho]. \tag{9}$$

In a similar manner, we have $V[O]_\rho = \mathrm{tr}[(O - E[O]_\rho \mathbb{I})^2\rho] = \mathrm{tr}[O^2\rho] - \mathrm{tr}[O\rho]^2$. These formulae deserve to be remembered since they enable us to calculate these quantities using operator O and ρ directly, while one has to solve the eigenvalue problems of O in order to obtain probabilities according to Born's statistical formula.

For an observable O, we say ρ is a *deterministic state* of O if the probability distribution $(\mathrm{tr}[P_i\rho])_{i=1}^m$ is deterministic, i.e., there exists $i_0 \in \{1, \ldots, m\}$ such that $\mathrm{tr}[P_{i_0}\rho] = 1$, $\mathrm{tr}[P_i\rho] = 0$ $(i \neq i_0)$.

Exercise 4. Show that a pure state ρ_ψ is deterministic of an observable O if and only if ψ is an eigenvector of O. In general, ρ is a deterministic state of an observable $O = \sum_{i=1}^m \lambda_i P_i$ if and only if that the support of ρ is included in one of the eingenspaces of O: i.e., there exists $i_0 \in \{0, \ldots, m\}$ such that $\mathrm{supp}\,\rho \subset P_{i_0}\mathcal{H} = E_O(\lambda_{i_0})$.

We say ρ is a *most random state* of $O = \sum_{i=1}^m \lambda_i P_i$ if the probability distribution for these is the uniform distribution, i.e., $\mathrm{tr}[P_i\rho] = \frac{1}{m}$.

[Examples] One of the most important observables for a qubit are those described by Pauli operators:

$$\sigma_x = \begin{pmatrix} 0 & 1 \\ 1 & 0 \end{pmatrix}, \quad \sigma_y = \begin{pmatrix} 0 & -i \\ i & 0 \end{pmatrix}, \quad \sigma_z = \begin{pmatrix} 1 & 0 \\ 0 & -1 \end{pmatrix}. \tag{10}$$

One can easily check they are all Hermitian with eigenvalues $+1$ and -1 with the following properties:

$$[\sigma_x, \sigma_y] = 2i\sigma_z, \ [\sigma_y, \sigma_z] = 2i\sigma_x, \ [\sigma_z, \sigma_x] = 2i\sigma_y, \tag{11a}$$

$$[\sigma_x, \sigma_y]_+ = [\sigma_y, \sigma_z]_+ = [\sigma_z, \sigma_x]_+ = 0, \ \sigma_x^2 = \sigma_y^2 = \sigma_z^2 = \mathbb{I}, \tag{11b}$$

where $[A, B] := AB - BA$ and $[A, B]_+ := AB + BA$ are commutator and anticommutator between operators A and B. Relations (11) are compactly summarized as

$$[\sigma_i, \sigma_j] = 2i \sum_{k=1}^3 \epsilon_{ijk}\sigma_k, \ [\sigma_i, \sigma_j]_+ = 2\delta_{ij} \ (\forall i, j = 1, 2, 3), \tag{12}$$

where $\sigma_1 = \sigma_x, \sigma_2 = \sigma_y, \sigma_3 = \sigma_z$, and ϵ_{ijk} and δ_{ij} denote the Levi-Civita symbol[6] and the Kronecker delta symbol[7], respectively. The eigenvectors of σ_z with eigenvalues $+1$ and -1 are given by $\psi_\uparrow^z := (1, 0)^T$ and $\psi_\downarrow^z := (0, 1)^T$, respectively. Then, it is easy to see that eigenvectors of σ_x are given by $\psi_\uparrow^x = \frac{1}{\sqrt{2}}(\psi_\uparrow^z + \psi_\downarrow^z)$ and $\psi_\downarrow^x = \frac{1}{\sqrt{2}}(\psi_\uparrow^z - \psi_\downarrow^z)$ with the corresponding eigenvalues $+1$ and -1.

Exercise 5. (I) Using (5), show that the probability distribution for observable σ_z under (pure) state of $\psi_\uparrow^z$ or $\psi_\downarrow^z$ is deterministic. (II) Show that the probability

---

[6] $\epsilon_{ijk} = 1$ if (ijk) is an even permutation of $(1, 2, 3)$; $\epsilon_{ijk} = -1$ if (ijk) is an odd permutation of $(1, 2, 3)$; otherwise $\epsilon_{ijk} = 0$.

[7] $\delta_{ii} = 1, \delta_{ij} = 0$ $(i \neq j)$.

distribution for observable σ_z under (pure) state $\psi_\uparrow^x$ or $\psi_\downarrow^x$ is uniform distribution, i.e., most random state of σ_z. (III) Show that the similar statement holds when interchanging x and z in (I) and (II). Moreover, show the similar relation between σ_x and σ_y, and also σ_y and σ_z. (IV) Using (9), calculate $E[\sigma_x]_{\psi_\uparrow^z}$.

Although there might be several preparations of a state using different procedures in experiments, we usually identify two states when there are no physical differences for any observables. This is physically a natural definition of the identification of states; otherwise there are no physical ways to see the difference of such states[*8]. To be precise, the density operator represents the equivalence class of (procedures of preparing) states with which probability distributions for any observables are identical. Indeed, it follows that for two density operators ρ_1, ρ_2, we have $\rho_1 \neq \rho_2$ (as operators) if and only if there exists observables with which the probability distributions for ρ_1 and ρ_2 are different.

Exercise 6. Show this fact using Born's statistical formula.

More compactly, one can show that $\rho_1 \neq \rho_2$ if and only if there exists an observable O such that $E[O]_{\rho_1} \neq E[O]_{\rho_2}$ [*9]. Therefore, henceforth, the expectation values are enough to identify or consider a statistical property of states.

Notice that for any set of density operators $\{\rho_i\}_{i=1}^n$ and any probability distribution $\{p_i\}_{i=1}^n$, the operator of convex combinations of them, i.e., $\rho := \sum_{i=1}^n p_i \rho_i$, is a density operator.

Exercise 7. Show this.

One of the preparation (or interpretation) of state $\rho = \sum_{i=1}^n p_i \rho_i$ is to prepare state ρ_i with probability p_i as an ensemble. Indeed, for any observable O, the expectation value in such ensemble should be $\sum_i p_i E[O]_{\rho_i}$, which is equal to $\mathrm{tr}[O\rho]$ using (9) and $\rho = \sum_{i=1}^n p_i \rho_i$. Since a density operator ρ is a positive operator with unit trace, it has an eigenvalue decomposition $\rho = \sum_{i=1}^d p_i |\phi_i\rangle\langle\phi_i|$ with eigenvalues being a probability distribution (see Exe. 1). Therefore, any density operator has a preparation by means of pure states; i.e., a preparation of $\rho = \sum_{i=1}^d p_i |\phi_i\rangle\langle\phi_i|$ is to prepare pure eigenstates ϕ_i with probability p_i. Notice that one should not adhere to one preparation for a density operator, since as mentioned above, density operators just represent an equivalence class among statistically equivalent preparations. To confirm this, consider the following two different preparation of an ensemble in qubit system: (i) prepare $\psi_{\uparrow}^z$ with probability $1/2$ and $\psi_{\downarrow}^z$ with probability $1/2$; (ii) prepare $\psi_{\uparrow}^x$ with probability $1/2$ and $\psi_{\downarrow}^x$ with probability $1/2$. Then, the density operator for (i) is $\rho_{(i)} = \frac{1}{2}|\psi_{\uparrow}^z\rangle\langle\psi_{\uparrow}^z| + \frac{1}{2}|\psi_{\downarrow}^z\rangle\langle\psi_{\downarrow}^z|$, and the density operator for (ii) is $\rho_{(ii)} = \frac{1}{2}|\psi_{\uparrow}^x\rangle\langle\psi_{\uparrow}^x| + \frac{1}{2}|\psi_{\downarrow}^x\rangle\langle\psi_{\downarrow}^x|$. However, since both $\{\psi_{\uparrow}^z, \psi_{\downarrow}^z\}$ and $\{\psi_{\uparrow}^x, \psi_{\downarrow}^x\}$ are orthonormal basis of $\mathbb{C}^2$, we have $\rho_{(i)} = \frac{1}{2}\mathbb{I} = \rho_{(ii)}$. Thus, these different preparations of states turn out to be (statistically) equivalent. As is later shown, a pure state is special in the sense that it does not have any preparation as an ensemble of different density operators with some probability, while any mixed state has such a preparation.

We have characterized quantum state as density operators describing pure and mixed states. However, since any density operator ρ has an eigenvalue decomposition $\rho = \sum_{i=1}^{d} p_i|\phi_i\rangle\langle\phi_i|$, it can be always interpreted as a preparation of pure states ϕ_i with probability p_i. Therefore, a mixed state can be always understood as a mixture of pure states. One might think that it is described by a mixed state just because we do not have an information on which pure states are indeed prepared[*10]. However, as is later shown, there are another origin of mixture from correlations between the system of interest and the environment. This makes a description of quantum states by density operators including not only pure state but also mixed states to be indispensable.

If two observables O_1 and O_2 are commutative, i.e., $[O_1, O_2] = 0$, there exists the common eigenvector ψ for both O_1 and O_2, since commutative Hermitian operators can be simultaneously diagonalized. Therefore, there exists a (pure) state ρ_ψ which is simultaneously deterministic of O_1 and O_2. In this regard, commuting observables are said to be **compatible observables**. However, two non-commutative observables O_1 and O_2 do not in general have a common deterministic state. For instance, while only deterministic pure states of σ_z are $\psi_{\uparrow}^z$ or $\psi_{\downarrow}^z$ (see Exe. 4), neither of them are deterministic of σ_x (see Exe. 5)[*11]. These are one of other remarkable

---

[*10]This origin of mixture of states from pure states as a lack of information is known as the **proper mixture**, or the **first kind of mixture**.[17] This is indeed the same as mixed states (probability distribution in the phase space) in classical statistics.

[*11]Indeed, they are not only indeterministic of σ_x, but also most random state for σ_x, and vice versa. In general, observable O_1 and O_2 are said to be **complementary ob-**

features of quantum mechanics, which reflects one aspect of **uncertainty principle**. The famous relation which characterizes this fact is the **uncertainty relation of Robertson type**:

$$\sigma[X]_\rho \sigma[Y]_\rho \geq \frac{1}{2}|\operatorname{tr}([X,Y]\rho)|. \tag{13}$$

Exercise 8. Indeed, stronger relation[19,20] is known:

$$\sigma[X]_\rho \sigma[Y]_\rho \geq \frac{1}{2}\operatorname{tr}(|\sqrt{\rho}[X,Y]\sqrt{\rho}|). \tag{14}$$

Show (14) and (13).

It is sometimes argued that these uncertainty relations imply that generally two observables cannot be simultaneously measured. Indeed, it is considered that two non-commuting observables cannot be simultaneously measured, which is another remarkable feature of quantum mechanics. Therefore, quantum mechanics does not generally provide a joint probability distribution among different observables. However, for compatible observables X and Y, one can simultaneously measure them and the joint probability distribution is given by

$$\Pr\{x_i, y_j\|X, Y, \rho\} = \operatorname{tr}(P_i^X P_j^Y \rho) = \operatorname{tr}(P_j^Y P_i^X \rho), \tag{15}$$

where $\{x_i\}$ ($\{y_j\}$) is the set of distinct eigenvalues of X (Y) with the corresponding eigenprojection P_i^X (P_j^Y).

Exercise 9. Prove that $\operatorname{tr}(P_i^X P_j^Y \rho)$ in (15) is mathematically a joint probability distribution, i.e., $\operatorname{tr}(P_i^X P_j^Y \rho) \geq 0, \sum_{i,j} \operatorname{tr}(P_i^X P_j^Y \rho) = 1$.

It should be noticed that this fact is consistently obtained from Born's statistical formula for a single observable as follows. First, notice that any commuting Hermitian operators X and Y are simultaneously diagonalizable, and thus there exists observable C, and functions $f, g : \mathbb{R} \to \mathbb{R}$ such that $X = f(C)$ and $Y = g(C)$.

---

servables if the deterministic state of O_1 is most random state for O_2, and vice versa. Observables σ_x and σ_z are a typical example of a complementary pair of observables. Similarly, the pair σ_x and σ_y, and also the pair σ_y and σ_z are complementary pairs of observables. Notice that two orthonormal base $\{\phi_i\}_{i=1}^d, \{\psi_i\}_{i=1}^d$ are called mutually unbiased if they satisfy $|\langle\phi_i, \psi_j\rangle|^2 = \frac{1}{d}$. Therefore, two observables are complementary iff the orthonormal base for them are mutually unbiased. Pauli operators are a typical examples that they are mutually unbiased, or complementary.

12

Exercise 10. We physically identify two observable A and B if there are no statistical differences for any state. Using this, show that the measurement of $X = f(C)$ can be performed in the following sense: Perform the measurement of C and if the output of C is c, then conclude the output of X is given by $f(c)$.

Therefore, one can simultaneously measure X and Y by measuring C and providing $f(c)$ and $g(c)$ as the outputs of X and Y when observing output c for the measurement of $C = \sum_k c_k P_k^C$ (spectral decomposition). Through this consideration, the joint probability distribution of X and Y is given by

$$\Pr\{x_i, y_j || X, Y, \rho\} = \sum_{k \ s.t., \ f(c_k)=x_i, \ g(c_k)=y_j} \text{tr}[P_k^C \rho].$$

On the other hand, it follows $P_i^X = \sum_{k, f(c_k)=x_i} P_k^C$ and $P_j^Y = \sum_{k, g(c_k)=y_j} P_k^C$. Therefore, we have $\sum_{k \ s.t., \ f(c_k)=x_i, \ g(c_k)=y_j}(P_k^C) = P_i^X P_j^Y$, and obtain (15).

[**Geometrical Structure of the State Space**] Let $\mathcal{L}(\mathcal{H})$ and $\mathcal{L}(\mathcal{H})_h$ be the sets of all the linear operators and all the Hermitian operators on $\mathcal{H}$, respectively. We denote by $\mathcal{S}(\mathcal{H})$ the set of all the density operators, i.e., $\mathcal{S}(\mathcal{H}) := \{\rho \in \mathcal{L}(\mathcal{H}) \mid \rho \geq 0, \ \text{tr}\,\rho = 1\} \subset \mathcal{L}(\mathcal{H})_h$. Hereafter we call $\mathcal{S}(\mathcal{H})$ the *state space*. As a mathematical conclusion of Exe. 7, we find that $\mathcal{S}(\mathcal{H})$ is a convex subset of $\mathcal{L}(\mathcal{H})_h$. Moreover, one can prove that $\mathcal{S}(\mathcal{H})$ is a bounded closed convex set of $\mathcal{L}(\mathcal{H})_h$ with some metric, and thus it is compact in $\mathcal{L}(\mathcal{H})_h$. (Recall that we are in the finite dimensional Hilbert space .)

Exercise 11. Explain this in details using the Heine-Borel theorem for a Euclidean space[*12].

Here we show that $\mathcal{S}(\mathcal{H})$ is bounded and closed with respect to the trace norm (See Appendix A): For any ρ_1, ρ_2, it follows that $||\rho_1 - \rho_2||_1 \leq ||\rho_1||_1 + ||\rho_2||_1 = \text{tr}\,\rho_1 + \text{tr}\,\rho_2 = 2$, and thus $\mathcal{S}(\mathcal{H})$ is bounded. In order to see the closedness, let $\rho_n \in \mathcal{S}(\mathcal{H})$ be a convergent sequence to an element $\rho \in \mathcal{L}(\mathcal{H})_h$ with respect to the trace norm. Then, what we have to show is that $\rho \in \mathcal{S}(\mathcal{H})$, namely $\text{tr}\,\rho = 1$ and $\rho \geq 0$. Using the inequality $|\text{tr}\,A| \leq ||A||_1$, it is easy to see $\text{tr}\,\rho = 1$. For positivity, using the Schwartz inequality and the inequality $||\rho - \rho_n||_\infty \leq ||\rho - \rho_n||_1$, one can show that $\langle \psi, \rho_n \psi \rangle \to \langle \psi, \rho \psi \rangle$

[*12]On the other hand, in infinite dimensional systems, this is no longer valid (Show this). However, by using a physically natural topology (weak topology for ρ), one can still show that $\mathcal{S}(\mathcal{H})$ is compact.

as $n \to \infty$, and thus the positivity of ρ holds. Therefore, we have shown that $\mathcal{S}(\mathcal{H})$ is a bounded closed subset of $\mathcal{L}(\mathcal{H})_h$.

Exercise 12. Show that the set of $\mathcal{L}(\mathcal{H})_h$ is a real Banach space with, e.g., the trace norm, Hilbert-Schmidt norm, and the operator norm with the dimension d^2 (Prove this).

Important property of quantum states from both mathematical and physical view points is that $\rho \in \mathcal{S}(\mathcal{H})$ is a pure state if and only if it is an extreme point of $\mathcal{S}(\mathcal{H})$ (Remind that extreme point ρ of convex set $S(\mathcal{H})$ is a state with no nontrivial convex combinations: See Appendix A).

Exercise 13. Prove that the set of the extreme points of $\mathcal{S}(\mathcal{H})$ is the set of all the pure states.

Namely, pure states are states which can not be nontrivially decomposed by different states. This is one of the reasons why they are called pure. On the other hand, a mixed state has always a nontrivial decomposition of different states. The eigenvalue decomposition of $\rho = \sum_i \lambda_i |\psi_i\rangle\langle\psi_i|$ means that any quantum state can be described by a convex combination of extreme points. We have thus:

Theorem 1. *State space $\mathcal{S}(\mathcal{H})$ is a compact convex subset of $\mathcal{L}(\mathcal{H})_h$. The set of extreme points of $\mathcal{S}(\mathcal{H})$ is the set of all the pure state.*

[**Several characterization of pure and mixed states**] Indeed, there are several characterizations of pure states and mixed states both from mathematical and physical point of view. First of all, pure state $\rho = |\psi\rangle\langle\psi|$ has a form of eigenvalue decomposition, which means that it is a rank one operator; conversely, any rank 1 density operator ρ is pure because the sum of eigenvalues is 1; also ρ is a pure state if and only if the (distinct) eigenvalues of ρ are $\{0, 1\}$. One can also verify that $\rho^2 = \rho$, namely it is a projection operator, is another necessary and sufficient condition for ρ to be a pure state[*13].

As another physical property of the pure (and mixed) states, it holds that ρ is a pure state if and only if there exists an nondegenerate observable O such that the probability distribution $p_i := \Pr\{\lambda_i || O, \rho\}$ ($i = 1, \ldots, d$) is deterministic, i.e., there exists $i_0 \in \{1, \ldots, d\}$ such that $p_{i_0} = 1$ while $p_i = 0$ for any $i \neq i_0$.

[*13] To see these, just use the fact that any density operator ρ has the eigenvalue decomposition $\rho = \sum_i p_i |\phi_i\rangle\langle\phi_i|$ with $\{p_i\}$ being a probability distribution.

14

Exercise 14. Prove this.

It is easy to see that an operator $\rho_m := \frac{1}{d}\mathbb{I}$ is an density operator. This is called the **maximal mixed state** (Since $\rho_m^2 \neq \rho_m$, this is a mixed state). This is considered as the most random state, or state with most uncertainty, since this always gives a uniform probability distribution for any nondegenerate observable O, i.e., it holds $\Pr\{\lambda_i||O, \rho\} = \mathrm{tr}[\frac{1}{d}|\phi_i\rangle\langle\phi_i|] = \frac{1}{d}$.

Exercise 15. Show that such a state is unique, which is $\frac{1}{d}\mathbb{I}$.

The **purity** $P(\rho)$ for a density operator ρ is defined by

$$P_S(\rho) := \mathrm{tr}[\rho^2] = \sum_{i=1}^{d} p_i^2,$$

where p_i's are eigenvalues of ρ. This is a useful measure for "pureness" of the density operator ρ. Indeed, one can show that for any density operator ρ, it holds $\frac{1}{d} \leq P(\rho) \leq 1$; The lower equality holds if and only if ρ is the maximal mixed state, and the upper equality holds if and only if ρ is a pure state.

Exercise 16. Prove this.

Another measure for the "pureness" of ρ is given by the **von Neumann entropy**:

$$S(\rho) := -\mathrm{tr}[\rho\ln\rho],$$

where ln is the natural logarithm*[14]. To be precise, $S(\rho)$ is defined as $\mathrm{tr}[f(\rho)]$ where $f : [0,1] \to \mathbb{R}$ is the (concave) function defined by $f(x) = -x\ln x$ with a convention $0\ln 0 = 0$ (see [Function of an operator] in Appendix A).

From the eigenvalue decomposition of ρ, one can show $S(\rho)$ is the Shannon entropy for the probability distribution of eigenvalues of ρ:

$$S(\rho) = H(\{p_i\}) := -\sum_{i=1}^{d} p_i\ln p_i,$$

where H is the Shannon entropy.

---

*[14]Here, we adopt the natural logarithm, but any choice of basis, usually 2 (bib), is not relevant for the later discussion.

More generally, one can define quantum **Tsallis entropy**[21] with a parameter $q > 0$:

$$S_q(\rho) = H_q(\{p_i\}) := \frac{1}{1-q}(\sum_{i=1}^{d} p_i^q - 1), \tag{16}$$

where H_q is called a (classical) Tsallis entropy. This is a one parameter extension of the von Neumann entropy, satisfying $S(\rho) = \lim_{q \to 1} S_q(\rho)$. In particular, $S_2(\rho)$ is called the quantum **linear entropy**: $S_2(\rho) = 1 - \operatorname{tr} \rho^2 = 1 - P(S)$. Notice that quantum **Tsallis entropy** is positive and concave function of (p_i) (thus $P(\rho)$ is a convex function of (p_i)).

Exercise 17. Prove the positivity and the concavity of **Tsallis entropy**.

Moreover, one can show that $0 \leq S_q(\rho) \leq S_q(\frac{1}{d}\,\mathbb{I})$, where the lower equality holds if and only if ρ is a pure state, and the upper equality holds if and only if ρ is the maximal mixed state (Show this). Therefore, Tsallis entropy, including von Neumann entropy, can be used again as a measure of the "pureness" of ρ. We summarize the characterization of pure states as follows:

Proposition 1. *For a density operator $\rho \in \mathcal{S}(\mathcal{H})$, the followings are equivalent:*

(i) ρ is a pure state,

(ii) ρ is a rank 1 operator; or the set of (distinct) eigenvalues of ρ is $\{0,1\}$,

(iii) ρ is a projection operator, i.e., $\rho^2 = \rho$,

(iv) ρ is an extreme point of $\mathcal{S}(\mathcal{H})$,

(v) There exists an nondegenerate observable such that the probability distribution under ρ is deterministic,

(vi) The purity takes the maximum value: $P(\rho) = 1$,

(vii) Tsallis entropy takes the minimum value $S_q(\rho) = 0$.

[State tomography and Bloch vectors] Born's statistical formula enables us to calculate any probability for a given observable and state. However, it is not clear how one can determine a state when one does not know the state under consideration. This is because a state in quantum mechanics is represented by a density operator, which is not composed of experimental data, like expectation values, but of an abstract operator. In principle, the measurements for all the observables determine a unique state as we have seen above. However, it is desirable to find the minimum required set of observables to determine a state, preferably with an algorithm to determine the corresponding density operator. Here we provide a systematic method for this problem, which leads us another state-representation with a physical quantity.

First of all, the number of independent real parameter for a density operator in a d-level system is $d^2 - 1$ (Check this from the property of the density operator of Hermiticity and unit trace). If one treats expectation values as a fundamental physical data, then, the minimum number of observables will be $d^2 - 1$ by considering that one gets an expectation value from each observables[*15]. This can be reconfirmed as follows: Let $\{F_i = F_i^\dagger\}_{i=1}^{d^2}$ be an orthogonal basis of $\mathcal{L}(\mathcal{H})_h$ with respect to the Hilbert-Schmidt inner product $\langle A, B \rangle_{HS} := \text{tr}[AB]$. The convenient choice of a basis is the one which satisfies $\langle F_i, F_j \rangle_{HS} = 2\delta_{ij}$ $(i, j = 1, \ldots, d^2 - 1)$ and $F_{d^2} := \frac{1}{\sqrt{d}} \mathbb{I}$ [*16]. Since, the expansion of any $A \in \mathcal{L}(\mathcal{H})_h$ with respect to the basis reads $A = \frac{1}{d}(\text{tr}\, A)\mathbb{I} + \frac{1}{2} \sum_{j=1}^{d^2-1} \text{tr}[F_j A] F_j$, any density operator ρ can be written as

$$\rho = \frac{1}{d}\mathbb{I} + \frac{1}{2} \sum_{j=1}^{d^2-1} E[F_j]_\rho F_j,$$

where we used $\text{tr}\,\rho = 1$ and the formula for the expectation value $E[F_j]_\rho = \text{tr}(F_j\rho)$. Since the right hand side of the equation can be obtained from physical data, i.e., the expectation values $E[F_j]_\rho$, this means that what one has to measure are $d^2 - 1$ expectation values of F_j's in order to determine the corresponding density operator. Moreover, this tells us that $(d^2 - 1)$ dimensional real vectors in $\mathbb{R}^{d^2-1}$ with the components of expectation values of F_j's provide a representation of a quantum states, which is equivalent to the representation with respect to density operators. One of the merit of such representation is that a quantum state is expressed by physical data, which are directly available for experimentalists. For $d = 2$, Pauli operators (10) and $\frac{1}{\sqrt{2}}\mathbb{I}$ are those comprising the orthogonal basis, and the representation with respect to 3-dimensional real vectors whose components are the expectation values of Pauli operators is known as the Bloch vector representation.[22] For general d, see 23,24,26.

So far, we have fixed a quantum system. Now, we will explain the com-

---

[*15]Since from each observable, one obtains a $d - 1$ numbers of data from a probability distribution, and thus in this sense, the minimum required numbers of observables is $d + 1$.

[*16]We have thus $\text{tr}\, F_i = 0$ and $\text{tr}[F_i F_j] = \delta_{ij}$ for all $i, j = 1, \ldots, d^2 - 1$. In mathematics, F_i $(i = 1, \ldots, d^2 - 1)$ are orthogonal generators of $SU(d)$.

position of quantum systems.

[P4: Composite Systems] *The composite system $A + B$ is described by the tensor Hilbert space $\mathcal{H}_A \otimes \mathcal{H}_B$, where A and B quantum systems with associated Hilbert spaces $\mathcal{H}_A$ and $\mathcal{H}_B$, with dimensions d_A and d_B. Observable O_A of A in the composite system is described by a Hermitian operator $O_A \otimes \mathbb{I}_B$, where $\mathbb{I}_B$ is the identity operator on $\mathcal{H}_B$. Similarly, observable O_B in B is described by $\mathbb{I}_A \otimes O_B$ on $\mathcal{H}_A \otimes \mathcal{H}_B$.*

In the following, we denote the set of observables by $\mathcal{O}(\mathcal{H})$ for a system described by $\mathcal{H}$[17]. Note that, for any pair of observables $O_A \in \mathcal{O}(\mathcal{H}_A)$ and $O_B \in \mathcal{O}(\mathcal{H}_B)$, $O_A \otimes \mathbb{I}_B$ and $\mathbb{I}_A \otimes O_B$ are compatible observables, and thus can be simultaneously measured: Let $O_A = \sum_i a_i P_i^A$ and $O_B = \sum_j b_j P_j^B$ be the spectral decompositions of O_A and O_B. Then the spectral decompositions $O_A \otimes \mathbb{I}_B$ and $\mathbb{I}_A \otimes O_B$ are $O_A = \sum_i a_i (P_i^A \otimes \mathbb{I}_B)$ and $O_B = \sum_j b_i (\mathbb{I}_A \otimes P_j^B)$. Therefore, from (15), the joint probability distribution to simultaneously measure O_A and O_B under a composite state $\rho \in \mathcal{S}(\mathcal{H}_A \otimes \mathcal{H}_B)$ is given by

$$\Pr\{a_i, b_j || O_A \otimes \mathbb{I}_B, \mathbb{I}_A \otimes O_B, \rho\} = \operatorname{tr}[(P_i^A \otimes \mathbb{I}_B)(\mathbb{I}_A \otimes P_j^B)\rho] = \operatorname{tr}[(P_i^A \otimes P_j^B)\rho]. \tag{17}$$

[Correlations] One of the important information gained from composite systems $A + B$ is the correlation between them. We say there are **no correlations in state** $\rho \in \mathcal{S}(\mathcal{H}_A \otimes \mathcal{H}_B)$ if the joint probability distribution for any pair of observables $O_A \in \mathcal{O}(\mathcal{H}_A)$ and $O_B \in \mathcal{O}(\mathcal{H}_B)$ is the product of the probability distributions for $O_A \otimes \mathbb{I}_B$ and $\mathbb{I}_A \otimes O_B$ in ρ:

$$\Pr\{a_i, b_j || O_A \otimes \mathbb{I}_B, \mathbb{I}_A \otimes O_B, \rho\} = \Pr\{a_i || O_A \otimes \mathbb{I}_B, \rho\} \Pr\{b_j || \mathbb{I}_B \otimes O_B, \rho\}. \tag{18}$$

Otherwise, we say there are correlations in ρ: i.e., ρ is a state with correlations if there are a pair of observables which are statistically correlated.

It should be notice that for any $O_A \in \mathcal{O}(\mathcal{H}_A)$ and for any $O_B, O_B' \in \mathcal{O}(\mathcal{H}_B)$, it follows that[18]

$$\sum_j \Pr\{a_i, b_j || O_A \otimes \mathbb{I}_B, \mathbb{I}_A \otimes O_B, \rho\} = \sum_j \Pr\{a_i, b_j' || O_A \otimes \mathbb{I}_B, \mathbb{I}_A \otimes O_B', \rho\}. \tag{19}$$

This means that no statistical differences appears in system A between the choice of O_B and O_B' in system B without knowing the output b_j or b_j'.

---

[17]Namely, $\mathcal{O}(\mathcal{H})$ is nothing but $\mathcal{L}(\mathcal{H})_h$. We shall use the notation $O \in \mathcal{O}(\mathcal{H})$ to emphasize that O is not only Hermitian operator but also is considered as an observable.
[18]This follows from (17) and $\sum_j P_j^B = \mathbb{I}_B$.

Exercise 18. Explain that if this condition is violated, one can use an ensemble of ρ for a superluminal communication between remote systems A and B.

In this sense, the condition (19) is called the **no-signaling condition** and it is considered as one of the basic principle which should be satisfied for any healthy physical theory, especially for the peaceful coexistence with the theory of relativity.

[Reduced density operators] Notice that even if you are interested in a fixed quantum system, say S, your system might be influenced by the existence of an outer system E, called an **environment**. In this case, it is important to treat your system as an open systems, and treat the composite system $S + E$ including all the environment. Here we explain how to describe a local state S under an influence of the environment. Let $\mathcal{H}_S$ and $\mathcal{H}_E$ be the Hilbert spaces for system S and environment E with $\mathcal{H}_{tot} = \mathcal{H}_S \otimes \mathcal{H}_E$, and let $\rho \in \mathcal{S}(\mathcal{H}_{tot})$ be the total state. Suppose that you are only interested in the system S, or suppose that you can only touch on the system S [19]. Namely, what you (can) measure are restricted to observables only in system S. In this situation, the state of S is described by the so-called **reduced density operator** $\rho_S := \mathrm{tr}_E \, \rho$, where tr_E is the partial trace [20] over the environment.

Exercise 19. Prove that mathematically the reduced density operator is a density operator in $\mathcal{H}_S$.

This is because what you measure are only local observables $O_S \in \mathcal{O}(\mathcal{H}_S)$. Indeed, since the total state is in $\rho \in \mathcal{S}(\mathcal{H}_S \otimes \mathcal{H}_E)$, the expectation value of such local observables are given by $\mathrm{tr}_{SE}(O_S \otimes \mathbb{I}_E \, \rho)$, and it follows

$$\mathrm{tr}_{SE}(O_S \otimes \mathbb{I}_E \, \rho) = \mathrm{tr}_S(O_S \, \mathrm{tr}_E \, \rho) = \mathrm{tr}_S(O_S \rho_S).$$

For those who are interested only the system S, it is consistent to think the system is in the reduced state ρ_S. Similarly, the reduced state ρ_E is

---

[19] It is even probable that you do not know the existence of environment.

[20] For an operator of the form $A \otimes B$ where A and B are operator on $\mathcal{H}_S$ and $\mathcal{H}_E$, $\mathrm{tr}_E \, A \otimes B := (\mathrm{tr}_E \, B)A$. Then, tr_E for any operator on $\mathcal{H}_S \otimes \mathcal{H}_E$ is defined by its linear extension. Equivalently it can be defined as follows: let $\{e_k\}_k$ be an orthonormal basis of $\mathcal{H}_E$; then the partial trace tr_E is a linear operator from $\mathcal{L}(\mathcal{H}_S \otimes \mathcal{H}_E)$ to $\mathcal{L}(\mathcal{H}_S)$ such that for any $X \in \mathcal{L}(\mathcal{H}_S \otimes \mathcal{H}_E)$, it satisfies

$$\langle \chi, \mathrm{tr}_E[X]\psi \rangle = \sum_k \langle \chi \otimes e_k, X\psi \otimes e_k \rangle,$$

for any $\chi, \psi \in \mathcal{H}_S$.

defined by taking partial trace over system S, which describes the state of the environment.

Exercise 20. Show that ρ is a state with no correlations if and only if $\rho = \rho_S \otimes \rho_E$, where ρ_S and ρ_E are reduced density operators of ρ.

Exercise 21. [Schmidt decomposition] For any state vector $\psi \in \mathcal{H}_A \otimes \mathcal{H}_B$, show that there exists orthonormal bases $\{\phi_i^A\}_{i=1}^{d_A}$ and $\{\phi_j^B\}_{j=1}^{d_B}$ of $\mathcal{H}_A$ and $\mathcal{H}_B$ such that

$$\psi = \sum_{i=1}^{l} \sqrt{\lambda_i} \phi_i^A \otimes \phi_i^B, \tag{20}$$

where $l \leq \min[d_A, d_B]$, and $\lambda_i > 0$, $\sum_{i=1}^{l} \lambda_i = 1$.

This is known as **Schmidt decomposition** of ψ, where l and λ_i's are called the **Schmidt number** and **Schmidt coefficients**, respectively. The reduced density operator of $|\psi\rangle\langle\psi|$ is

$$\rho_A = \mathrm{tr}_B(\sum_{i,j} \sqrt{\lambda_i \lambda_j} |\phi_i^A\rangle\langle\phi_j^A| \otimes |\phi_i^B\rangle\langle\phi_j^B|) = \sum_{i=1}^{l} \lambda_i |\phi_i^A\rangle\langle\phi_i^A|, \tag{21}$$

$\mathrm{tr}_B |\phi_i^B\rangle\langle\phi_j^B| = \delta_{ij}$. This shows that the Schmidt number, Schmidt coefficients and the orthonormal bases in the Schmidt decomposition are nothing but the rank, eigenvalues, and orthonormal eigenvectors of reduced states, respectively.

Exercise 22. As a corollary of Schmidt decomposition, show that for any density operator $\rho_A \in \mathcal{H}_A$, there exists a state vector $\psi \in \mathcal{H}_A \otimes \mathcal{H}_R$ with some ancillary Hilbert space $\mathcal{H}_R$ such that ρ_A is the reduced state of $|\psi\rangle\langle\psi|$.

This is called a **purification** of ρ_A and implies another origin of mixture of states, which is important to understand the mechanics of decoherence. Indeed, *even though the total system is in pure state, the reduced state is in general a mixed state*. For instance, in two qubits system, let $\{\psi_i^S \in \mathcal{H}_S\}_{i=1,2}$ and $\{\psi_i^E \in \mathcal{H}_E\}_{i=1,2}$ be orthonormal bases for $\mathcal{H}_S$ and $\mathcal{H}_E$, respectively, and let

$$\psi = \frac{1}{\sqrt{2}} \left(\psi_1^S \otimes \psi_1^E + \psi_2^S \otimes \psi_2^E \right). \tag{22}$$

This has already the form of Schmidt decomposition, and the reduced density operator for S is thus

$$\mathrm{tr}_E |\psi\rangle\langle\psi| = \frac{1}{2}(|\psi_1^S\rangle\langle\psi_1^S| + |\psi_2^S\rangle\langle\psi_2^S|) = \frac{1}{2}\mathbb{I}_S, \tag{23}$$

which is the maximally mixed state of S. In this way, a mixed state is required by considering the influence of environment. More generally, we have the following statement.

Theorem 2. *If a total density operator $\rho \in \mathcal{S}(\mathcal{H}_S \otimes \mathcal{H}_E)$ has (nonzero) correlations between S and E, then the reduced state ρ_S is a mixed state.*

To show this, let us first consider the case where the total state ρ is pure with the Schmidt decomposition (20) and the reduced density operator (21); if the reduced density operator is pure, i.e., it is a rank 1 operator, then the Schmidt number $l = 1$, and thus the state vector has the decomposition $\psi = |\psi_1^S\rangle\langle\psi_1^S| \otimes |\psi_1^E\rangle\langle\psi_1^E| = |\psi_1^S \otimes \psi_1^E\rangle\langle\psi_1^S \otimes \psi_1^E|$. Therefore, from Exe. 20, ρ is a state with no correlations.

Exercise 23. Verify Theorem 2 for a general case where the total state ρ is mixed. [Hint: use a purification ρ into $(\mathcal{H}_S \otimes \mathcal{H}_E) \otimes \mathcal{H}_R$ with some (hypothetical) ancilla $\mathcal{H}_R$].

Importantly, a physical meaning of Theorem 2 is that correlations with an environment could be the origin of the mixture of the reduced state[*21].

Finally, we explain the dynamics of quantum mechanics; Indeed, once the state space is established, then the dynamics is introduced as a trajectory on the state space with a single parameter, namely a time:

[P5: Dynamics for isolated quantum systems] *If quantum system S is (dynamically) isolated from the environment[*22], the time evolution of the state is governed by the **von Neumann-Schrödinger equation**[*23]:*

$$\frac{d}{dt}\rho_t = -i[H, \rho_t], \tag{24}$$

*where ρ_t is a density operator at time t, and $H = H^\dagger$ is the **Hamiltonian** of the system.*

[*21] This origin of mixture is known as the **improper mixture** or the **second kind of mixture**.[17] The statement is quite general, and one can verify this not only in quantum systems, but also in any generic probability models with no-signaling condition.

[*22] Here, isolated system means a system which does not interact with the environment; the dynamics for a non-isolated system, i.e., for an open system, one should include any environment which interact with the system so that the total system is isolated. Notice that even an isolated system could have non-zero correlations with some environment.

[*23] Time derivative is usually defined by a trace norm including infinite dimensional cases; but since we are in a finite scheme, any norms work equivalently to define the derivative.

Exercise 24. Show that equation (24) is equivalent to

$$\rho_t = \mathcal{U}_t \rho := U_t \rho_0 U_t^\dagger, \tag{25}$$

where $U_t = \exp(-iHt)$ is a unitary operator, known as a **time-evolution operator**.

Therefore, one can conclude that the primary property of the dynamics for isolated quantum systems is the unitarity of the time evolution*24.

It is sometimes useful to consider a time-evolution map $\mathcal{U} : \mathcal{S}(\mathcal{H}) \to \mathcal{S}(\mathcal{H})$, which maps an initial state ρ to a state ρ_t, forgetting the parameter time t. This is useful when one's interest is a transition from an initial state to a final state, where the process between them is not relevant. Then, the unitary time evolution is given by the unitary map $\mathcal{U}\rho = U\rho U^\dagger$ with a unitary operator U on $\mathcal{H}$.

Notice that with any unitary time evolution $\mathcal{U}\rho = U\rho U^\dagger$, the purity does not change

$$\operatorname{tr}(\mathcal{U}\rho)^2 = \operatorname{tr}(U\rho U^\dagger)^2 = \operatorname{tr}(U\rho U^\dagger U\rho U^\dagger) = \operatorname{tr} U\rho^2 U^\dagger = \operatorname{tr}\rho^2, \tag{26}$$

where in the final step the cyclic property of the trace has been used.

On the other hand, in the real experiments, there are several phenomena where the purity of the system changes, such as transition to a thermal equilibrium state or, decoherence or dephasing process in quantum measurement. Therefore, in order to explain these phenomena without changing the basic law of quantum mechanics, it is natural to consider open quantum systems, which are not dynamically isolated with the environment. The phenomena, which cannot be described by a unitary time-evolution, are known as **decoherence**, the origin of which is usually considered due to the interaction with the environment.

[**POVM**] Although we have explained that any observable can be described by an Hermitian operator, this does not mean they describe all the possible measurements on the system. Indeed, instead of measuring an observable in S, one can first interact the system with some another system A (apparatus, environment, ancilla, etc. whatever you call), and then measure an observable on A to get some information on the system: More precisely, let system S be in a state $\rho \in \mathcal{S}(\mathcal{H})$. Let the apparatus A be described by

---

*24Mathematically, $\{U_t = \exp(-iHt)\}_{t\in\mathbb{R}}$ is a strongly continuous one-parameter group on the Hilbert space $\mathcal{H}$. Conversely, if $\{U_t\}_{t\in\mathbb{R}}$ is a strongly continuous one-parameter group on $\mathcal{H}$, then one can show that there exists the generator iH such that $U_t = \exp(-iHt)$ (Stone's Theorem).

Hilbert space $\mathcal{H}_E$, which is initially in a state $\sigma \in \mathcal{S}(\mathcal{H}_E)$ where the total state is described by $\rho \otimes \sigma \in \mathcal{S}(\mathcal{H}_S \otimes \mathcal{H}_E)$. Let U be a time evolution unitary operator on $\mathcal{H}_S \otimes \mathcal{H}_E$ with which systems S and A interact, thereby the final state is in $U\rho \otimes \sigma U^\dagger$:

$$\rho \otimes \sigma \mapsto U\rho \otimes \sigma U^\dagger.$$

Finally, let us perform a measurement of an observable $M = \sum_j m_j Q_j$ on A. According to the Born's statistical formula, the probability to obtain m_j is given by

$$\Pr\{m_j \,\|\, \mathbb{I} \otimes M, U\rho \otimes \sigma U^\dagger\} = \text{tr}_{SE}[\mathbb{I}_S \otimes Q_j U\rho \otimes \sigma U^\dagger].$$

This can be written, by defining $E_j := \text{tr}_E[\mathbb{I}_S \otimes \sigma U^\dagger \mathbb{I}_S \otimes Q_j U] \in \mathcal{L}(\mathcal{H}_S)$, as

$$\Pr\{m_j \,\|\, \mathbb{I} \otimes M, U\rho \otimes \sigma U^\dagger\} = \text{tr}_S[E_j \rho], \qquad (27)$$

using the cyclic property of the trace. Since this whole process reflects some information on the system S by the effect of interaction U, this is considered a kind of measurement on S which is called an **indirect measurement**.

Exercise 25. Prove that the operator-set $\{E_j := \text{tr}_A[\mathbb{I}_S \otimes \sigma U^\dagger \mathbb{I}_S \otimes Q_j U] \in \mathcal{L}(\mathcal{H}_S)\}$ satisfies

$$E_j \geq 0, \quad \sum_j E_j = \mathbb{I}_S. \qquad (28)$$

Importantly, the converse is also satisfied:

Theorem 3. *The measurement of any set of operators $\{E_j\}$ on $\mathcal{H}_S$ satisfying (28) can be realized by an indirect measurement. Namely, there exists a Hilbert space $\mathcal{H}_E$, a state $\sigma \in \mathcal{S}(\mathcal{H}_E)$, a unitary operator $U \in \mathcal{O}(\mathcal{H}_S \otimes \mathcal{H}_E)$, and an observable $M = \sum_j m_j Q_j$ such that the probability to obtain the output m_j under a state $\rho \in \mathcal{S}(\mathcal{H}_S)$ is given by*

$$\text{tr}_S[E_j \rho] = \text{tr}_{SE}[(\mathbb{I}_S \otimes Q_j) U\rho \otimes \sigma U^\dagger]. \qquad (29)$$

To prove this, first show the following:

Exercise 26. Let $\{V_k\}_{k=1}^{d'}$ be a set of operators on $\mathcal{H}$ satisfying $\sum_{k=1}^{d'} V_k^\dagger V_k = \mathbb{I}$. Let $\mathcal{H}_E$ be a Hilbert space with dimension d' and let $\{e_k\}_{k=1}^{d'}$ be an orthonormal basis of $\mathcal{H}_E$. Then, there exists a Unitary operator U on $\mathcal{H}_S \otimes \mathcal{H}_E$ such that

$$U(\psi \otimes e_1) := \sum_{k=1}^{d'} (V_k \psi) \otimes e_k,$$

for any $\psi \in \mathcal{H}_S$. [Hint: First check an operator U defined on the subspace $\{\psi \otimes e_1 \,|\, \psi \in \mathcal{H}_S\} \subset \mathcal{H} \otimes \mathcal{H}_E$ preserves an inner product. Then, show it can be extended to the whole space keeping the unitarity.]

Proof of Theorem 3 Let $\{E_k\}_{k=1}^{d'}$ satisfy (28), and let $\mathcal{H}_E$ be a Hilbert space with dimension d' and let $\{e_k\}_{k=1}^{d'}$ be an orthonormal basis of $\mathcal{H}_E$. From Exe. 26, there exists a unitary operator on $\mathcal{H}_S \otimes \mathcal{H}_E$ such that

$$U(\psi \otimes e_1) := \sum_{k=1}^{d'} (\sqrt{E_k}\psi) \otimes e_k,$$

since $\mathbb{I} = \sum_k E_k = \sum_j \sqrt{E_k}^\dagger \sqrt{E_k}$. Notice that for any $\chi \in \mathcal{H}_S$, it follows

$$\langle \chi \otimes e_i, U(\psi \otimes e_1) \rangle = \langle \chi, \sqrt{E_i}\psi \rangle.$$

Let $\sigma := |e_0\rangle\langle e_0| \in \mathcal{S}(\mathcal{H}_E)$, and let $M := \sum_j m_j Q_j$ with $Q_j := |e_j\rangle\langle e_j|$ be an observable of $\mathcal{H}_E$. Then, for any $\chi \in \mathcal{H}$, we have

$$\mathrm{tr}_{SE}[(\mathbb{I}_S \otimes Q_j)U(|\chi\rangle\langle\chi| \otimes \sigma)U^\dagger] = \mathrm{tr}_{SE}[(\mathbb{I}_S \otimes |e_j\rangle\langle e_j|)U(|\chi \otimes e_0\rangle\langle\chi \otimes e_0|)U^\dagger]$$

$$= \sum_{i,k} \langle f_i \otimes e_k, (\mathbb{I}_S \otimes |e_j\rangle\langle e_j|)U(\chi \otimes e_0)\rangle\langle U(\chi \otimes e_0), f_i \otimes e_k\rangle$$

$$= \sum_i \langle f_i \otimes e_j, U(\chi \otimes e_0)\rangle\langle U(\chi \otimes e_0), f_i \otimes e_j\rangle = \sum_i \langle f_i, \sqrt{E_i}\chi\rangle\langle\sqrt{E_i}\chi, f_i\rangle$$

$$= \mathrm{tr}_S[|\sqrt{E_i}\chi\rangle\langle\sqrt{E_i}\chi|] = \mathrm{tr}_S[E_i|\chi\rangle\langle\chi|], \tag{30}$$

where $\{f_i\}$ is a orthonormal basis of $\mathcal{H}_S$. Since any density operator $\rho \in \mathcal{S}(\mathcal{H}_S)$ is a positive linear combination of the form $|\chi\rangle\langle\chi|$, we obtain (29). $\blacksquare$

Thus, any operator-set $\{E_j\}$ satisfying (28) can be considered as a measurement performed by means of an indirect measurement to obtain an outcome m_j with the probability $\mathrm{tr}_S[E_j\rho]$ under the state $\rho \in \mathcal{S}(\mathcal{H}_S)$. The set of $\{E_j\}$ with $E_j \geq 0$ and $\sum_j E_j = \mathbb{I}$ is known as a **positive operator valued measure**, or **probability operator valued measure** (hereafter **POVM**): POVM is considered as a possible measurement by which one obtains some output m_j with the probability $\mathrm{tr}[E_j\rho]$ under the state ρ.

The measurement of observable $O = \sum_i \lambda_i P_i$ is interpreted as a special class of POVM, since $P_i \geq 0$ and $\sum_i P_i = \mathbb{I}_S$. In particular, the set $\{P_i\}$ is called a **projection valued measure**.

Indeed, by only assuming that any quantum state is represented by a density operator, one can show that POVM measurement is the most general measurement, which is consistent with the interpretation of probability:[30,31] Phenomenologically, a measurement is a kind of action to obtain some outcome m_j from a system S. Suppose that the outcome is obtained with a certain probability $\mathrm{Pr}\{m_j \parallel \rho\}$, which depends on the state ρ. Then, the probability as a map from ρ to $\mathrm{Pr}\{m_j \parallel \rho\}$ should be affine, i.e., $\mathrm{Pr}\{m_j \parallel \lambda\rho_1 + (1 - \lambda)\rho_2\} = \lambda\mathrm{Pr}\{m_j \parallel \rho_1\} + (1 - \lambda)\mathrm{Pr}\{m_j \parallel \rho_2\}$ for any $\lambda \in [0,1], \rho_1, \rho_2 \in \mathcal{S}(\mathcal{H})$. This follows from that noting one of the preparation of the sate $\rho = \lambda\rho_1 + (1 - \lambda)\rho_2$ is to prepare states ρ_1 and ρ_2 with probabilities λ and $1 - \lambda$, respectively (Explain this in details). Then, it is not difficult to show that there exists a POVM $\{E_j\}_{j=1}^d$ such that $\mathrm{Pr}\{m_j \parallel \rho\} = \mathrm{tr}(E_j\rho)$.

Exercise 27. Prove the final part.

[**Measurement Process**] Finally, a comment on the measurement process is given. So far, we have only explained a statistical property of quantum mechanics under a given state, but we have not yet explained a description of a post-measurement state, namely a state after measurement. Indeed, another distinguished feature of quantum theory is that, contrary to the classical physics, one cannot perform a measurement without disturbing the system. The most well known example is the so-called as the uncertainty principle of the Heisenberg type.[27-29] Although there are still debates on the measurement problem in quantum theory, especially in association with an interpretation of the theory, the description of the measurement process is fairly established, at least in the operational or practical point of view. The full description of the measurement process is considered to be described by a completely positive instrument (completely positive map valued measure),[13,32,33] and it is known as a **measurement operator formalism** for discrete outputs in the field of quantum information theory.[1] We do not explain them here in details, but in order to motivate the decoherence problem, let us see a typical measurement process under the **repeatability hypothesis**.[34] Here, the repeatability hypothesis assumes that, after observing an output λ_{i_0} through a measurement, one will again obtain the same outcome λ_{i_0} with probability 1 if one performs the same measurement immediately after the first measurement. For the case of observable $O = \sum_i \lambda_i P_i$, however, such state ρ is restricted to be a density operator satisfying $\mathrm{supp}\,\rho \subset P_{i_0}\mathcal{H} = E_O(\lambda_{i_0})$ (see Exe. 4), irrespective of an initial state before measurement. For instance, it is uniquely determined to be ϕ_{i_0} for a nondegenerate observable $O = \sum_{j=1}^d \lambda_j |\psi_j\rangle\langle\psi_j|$. Thus, starting from an initial density operator ρ, a post-measurement state is $|\phi_{i_0}\rangle\langle\phi_{i_0}|$ if one obtains an outcome λ_{i_0} by a measurement of a nondegenerate observable. This is known as **von Neumann's projective measurement** [*25].

Next suppose that one performs the von Neumann's projective measurement of nondegenerate observable under a state ρ, but **does not know which output was obtained** in the measurement. Then, the description of the post measurement state should be by means of ensemble such that the state is in $|\psi_j\rangle\langle\psi_j|$ with probability $p_j = \text{tr}[|\psi_j\rangle\langle\psi_j|\rho] = \langle\psi_j, \rho\psi_j\rangle$. Consequently, the corresponding post-measurement state is represented by the following density operator:

$$\rho_{pm} = \sum_{j=1}^{d} \langle\psi_j, \rho\psi_j\rangle |\psi_j\rangle\langle\psi_j|.$$

This is in general mixed state, even when the initial state ρ is in pure state. Therefore, in general, during the measurement process from an initial state to the final state, the purity changes (in most cases decreases), and thus this cannot be explained by a unitary time-evolution (see Eq. (26)). This purity change is one of the typical phenomena of decoherence, which must be explored as an open quantum mechanics.

As another easy way to see the disturbance of measurement, suppose that you would like to distinguish two nonorthogonal pure states ψ_1 or ψ_2 in $\mathcal{H}_S$. Then it is easy to see that you cannot get any information for this purpose without causing any disturbance. To do this, you should prepare a reference state ϕ in the ancilla system $\mathcal{H}_A$, and by interacting with some unitary operator U, you want to get some information about ψ_1 or ψ_2: If the state is in ψ_i $(i = 1, 2)$ in $\mathcal{H}_S$, then

$$|\psi_i \otimes \phi\rangle\langle\psi_i \otimes \phi| \mapsto U(|\psi_i \otimes \phi\rangle\langle\psi_i \otimes \phi|)U^\dagger \ (i = 1, 2).$$

Assume that you can get some information without causing any disturbance to the system S. This requires that the reduced state of S from $U(|\psi_i \otimes \phi\rangle\langle\psi_i \otimes \phi|)U^\dagger$ remains to be $|\psi_i\rangle\langle\psi_i|$, and therefore from Theorem 2, it follows that

$$U(|\psi_1 \otimes \phi\rangle\langle\psi_1 \otimes \phi|)U^\dagger = |\psi_1\rangle\langle\psi_1| \otimes |\phi_1\rangle\langle\phi_1| \qquad (31)$$

$$U(|\psi_2 \otimes \phi\rangle\langle\psi_2 \otimes \phi|)U^\dagger = |\psi_2\rangle\langle\psi_2| \otimes |\phi_2\rangle\langle\phi_2|, \qquad (32)$$

[*25] For a general degenerate observable $O = \sum_i \lambda_i P_i$, the state change: $\rho \rightarrow \sum_i P_i \rho P_i$ is known as a Lüders projective measurement. Notice that this is not determined only by the repeatable hypothesis.

with some state vectors ϕ_1, ϕ_2 in A. However, by multiplying each side of (31) and (32) and taking a trace over $S + A$ for them:

$$\mathrm{tr}[U(|\psi_2 \otimes \phi\rangle\langle\psi_2 \otimes \phi|)U^\dagger U(|\psi_1 \otimes \phi\rangle\langle\psi_1 \otimes \phi|)U^\dagger]$$
$$= \mathrm{tr}[|\psi_2\rangle\langle\psi_2| \otimes |\phi_2\rangle\langle\phi_2||\psi_1\rangle\langle\psi_1| \otimes |\phi_1\rangle\langle\phi_1|], \qquad (33)$$

one obtains

$$|\langle\psi_1, \psi_2\rangle|^2 = |\langle\psi_1, \psi_2\rangle|^2 |\langle\phi_1, \phi_2\rangle|^2.$$

Since we have assumed $\langle\psi_1, \psi_2\rangle \neq 0$, it follows $|\langle\phi_1, \phi_2\rangle| = 1$, and thus $|\phi_1\rangle\langle\phi_1| = |\phi_2\rangle\langle\phi_2|$ (Prove this using the equality condition of the Schwatz's inequality). This means that one could not get any information about which state ψ_1 or ψ_2 was prepared. In other words, to get some information about this, one has to disturb the system S.

Exercise 28. [No-cloning theorem] In the similar manner, prove that unknown quantum states cannot be perfectly cloned.

3. Phenomenological view of a dynamics in quantum systems

So far, we have explained the dynamics for only isolated quantum systems. Non-isolated system interacting with its environment is called an **open system**. Before going to the general study of open quantum systems based on the law of quantum mechanics explained in the preceding section, let us examine a general nature of possible dynamics from the phenomenological point of view. Since we have already seen that a general quantum state is represented by a density operator ρ on the associated Hilbert space $\mathcal{H}$, a time evolution should keep the state in the state space $\mathcal{S}(\mathcal{H})$. If one would like to consider a dynamics as a map Φ such that it maps an initial density operator ρ into the final density operator ρ', it turns out to be natural to assume that Φ is an affine map on $\mathcal{S}(\mathcal{H})$; Here **affine map** on $\mathcal{S}(\mathcal{H})$ is a map Φ which satisfies

$$\Phi(\rho) = \lambda\Phi(\rho_1) + (1 - \lambda)\Phi(\rho_2), \qquad (34)$$

for any $\rho_1, \rho_2 \in \mathcal{S}(\mathcal{H})$, $\lambda \in [0, 1]$; $\rho := \lambda\rho_1 + (1 - \lambda)\rho_2$.

First of all, we (implicitly) assume that we can prepare any density operator, and it evolves uniquely to a final state *irrespective of the preparation of the state*. From this, one can consider a time evolution map Φ from $\mathcal{S}(\mathcal{H})$ to $\mathcal{S}(\mathcal{H})$, whose domain is whole $\mathcal{S}(\mathcal{H})$. Next, since one of the preparation of state $\lambda\rho_1 + (1 - \lambda)\rho_2$ is to prepare ρ_1 with probability λ and ρ_2 with probability $1 - \lambda$, the state $\Phi(\rho)$ should satisfy

$$\Pr\{m_j\|\{E_j\}, \Phi(\rho)\} = \lambda\Pr\{m_j\|\{E_j\}, \Phi(\rho_1)\} + (1 - \lambda)\Pr\{m_j\|\{E_j\}, \Phi(\rho_2)\},$$

for any POVM measurement $\{E_j\}$ on $\mathcal{H}$. Since $\Pr\{m_j \| \{E_j\}, \Phi(\rho_1)\} = \text{tr}[E_j\rho]$, it follows

$$\text{tr}[E_j\Phi(\rho)] = \lambda\,\text{tr}[E_j\Phi(\rho_1)] + (1-\lambda)\,\text{tr}[E_j\Phi(\rho_2)] = \text{tr}[E_j(\lambda\Phi(\rho_1) + (1-\lambda)\Phi(\rho_2))].$$

Since this follows for arbitrary POVM element E_j, the affinity (34) follows.

Therefore, a general time-evolution map Φ should be an affine map on $\mathcal{S}(\mathcal{H})$. Mathematically, such map is uniquely extended to the linear map on $\mathcal{L}(\mathcal{H})$:

Exercise 29. Show that any $A \in \mathcal{L}(\mathcal{H})$ can be written as $\sum_{k=0}^{3} \alpha_k\rho_k$ with $\alpha_k \geq 0$, $\rho_k \in \mathcal{S}(\mathcal{H})$. Show that the representation is not unique, while each real and imaginary part is unique. [Hint: first, show that A can be uniquely decomposed with two Hermitian operators A_I, A_R such that $A = A_R + iA_I$. Next, show that any Hermitian operator B can be written as $B = C - D$ with two positive operators C and D. Notice that this decomposition is not unique (Explain this with some explicit examples).]

Exercise 30. Using Exe. 29, show that any affine map Φ on $\mathcal{S}(\mathcal{H})$ can be uniquely extended to the linear map on $\mathcal{L}(\mathcal{H})$. [Hint: Define the extension $\Phi(A) := \sum_{k=0}^{3} \alpha_k\Phi(\rho_k)$ and show the well-definedness by using the affinity of Φ, and the uniqueness of real and imaginary part of the decomposition. Finally, show that it is linear on $\mathcal{L}(\mathcal{H})$ and the extension is unique.]

Consequently, an affine map on $\mathcal{S}(\mathcal{H})$ is uniquely extended to be a **trace-preserving positive map** on $\mathcal{L}(\mathcal{H})$. In order to explain this, first we explain some mathematical ingredients for a linear map on $\mathcal{L}(\mathcal{H})$. Let $\Phi : \mathcal{L}(\mathcal{H}) \to \mathcal{L}(\mathcal{H})$ be a linear map on $\mathcal{L}(\mathcal{H})$.
[Trace Preserving Maps] Φ is called a trace-preserving map if Φ preserves the trace of any operator, i.e., $\text{tr}[\Phi(A)] = \text{tr}[A]$ for any $A \in \mathcal{L}(\mathcal{H})$.
[Positive Maps] Φ is called

(i) **Hermitian** if Φ preserves the Hermiticity of operators, i.e., $\Phi(A) \in \mathcal{L}(\mathcal{H})_h$ for any $A \in \mathcal{L}(\mathcal{H})_h$; and

(ii) **positive** if Φ preserves the positivity of operators: Namely, if $A \in \mathcal{L}(\mathcal{H})$ is a positive operator, then $\Phi(A)$ is a positive operator.
[Completely Positive Maps] There are stronger notions of positive maps: A linear map Φ is called

(iii) n-**positive** if $\Phi \otimes id_n$ is positive on $\mathcal{L}(\mathcal{H}) \otimes M_n(\mathbb{C})$, where id_n is an identity map on $\mathcal{L}(\mathbb{C}^n) \simeq M_n(\mathbb{C})$; and

(iv) **completely positive** if Φ is n-positive for all $n \in \mathbb{N}$.

Exercise 31. Show that 1-positivity is equivalent to just a positivity.

Exercise 32. Show that n-positivity implies m-positivity of Φ for any $m \leq n$.

Exercise 33. Show that positive map is an Hermitian map.

Exercise 34. One may consider n-Hermite map in the same manner as n-positive map. Show that n-Hermiticity is equivalent to just an Hermiticity.

Now one can easily show that the linearly extended map from an affine map on $\mathcal{S}(\mathcal{H})$ is a trace preserving positive map.

Exercise 35. Show this.

However, we should further restrict a possible time-evolution map by considering an existence of an outer world than a quantum system described by $\mathcal{H}$. Suppose that there exists a quantum system described by $\mathbb{C}^n$, and consider a total system described by $\mathcal{H} \otimes \mathbb{C}^n$. Although we are considering a general dynamics for a system $\mathcal{H}$ only, there should be a consistency to describe the dynamics from the point of view of total systems. Indeed, the prepared density operator on $\mathcal{H}$ might be a reduced state from a total state on $\mathcal{H} \otimes \mathbb{C}^n$. Suppose also that the system $\mathbb{C}^n$ does not interact with $\mathcal{H}$, then the time-evolution map corresponding to the map Φ on a local system $\mathcal{H}$ is described by $\Phi \otimes id_n$ where id_n is an identity map on $\mathcal{L}(\mathbb{C}^n)$. This require us that not only Φ but also the map $\Phi \otimes id_n$ should be positive, i.e., Φ should be n positive so that $\Phi \otimes id_n$ maps any density operator on $\mathcal{H} \otimes \mathbb{C}^n$ to a density operator there. Consequently, it is naturally required that *a general time evolution map Φ is a trace preserving and completely positive map* since n could be arbitrary.

Exercise 36. Show that a unitary map $\Phi_U(A) := UAU^\dagger$ with a unitary operator U on $\mathcal{H}$ is a TPCP map on $\mathcal{L}(\mathcal{H})$.

This shows that a general dynamics for an isolated quantum systems is (of course) consistent with our general view of dynamics.

One might worry whether there are other restrictions by considering different situations, e.g., the case with an interaction. However, we shall see in the next section that these are already enough as a physically tangible time evolution map for open quantum systems.

There are several useful equivalent characterization of completely positive maps (Hereafter CP maps).

Theorem 4. *Let $\mathcal{H}$ be a Hilbert space with dimension d and let Φ be a linear map on $\mathcal{L}(\mathcal{H})$. The following are equivalent:*

(i) *Φ is completely positive;*

(ii) *Φ is d-positive;*

(iii) *Let ψ_{me} be an (unnormalized) maximal entangled state on $\mathcal{H} \otimes \mathcal{H}$, i.e, $\psi_{me} = \sum_{i=1}^{d} \phi_i \otimes \phi_i$ where $\{\phi_i\}_{i=1}^{d}$ is an orthonormal basis of $\mathcal{H}$. Then, $\Phi \otimes id_d(|\psi_{me}\rangle\langle\psi_{me}|) \geq 0$;*

(iv) *There exists a set of operators $\{V_i \in \mathcal{L}(\mathcal{H})\}_{i=1}^{m}$ such that $\Phi(A) = \sum_{i=1}^{m} V_i A V_i^{\dagger}$.*

The property (ii) means that d-positivity is enough to show the completely positivity, in the case of d dimensional Hilbert space. Moreover, the property (iii) means that a maximally entangled state[*26] is enough to be checked among all positive operators in order to show positive preserving property of $\Phi \otimes id_d$. The representation of CP map in (iv) is sometimes called an **operator sum representation**.

[**Proof of Theorem 4**] Implications (i) $\Rightarrow$ (ii) $\Rightarrow$(iii) directly follow from definitions. Suppose (iii) holds. Since the operator $\sum_{ij} \Phi(E_{ij}) \otimes E_{ij} = \Phi \otimes id_d(|\psi_{me}\rangle\langle\psi_{me}|)$ is positive where $E_{ij} = |\phi_i\rangle\langle\phi_j|$, it has an eigenvalue decomposition

$$\sum_{ij} \Phi(E_{ij}) \otimes E_{ij} := \sum_{k} x_k |v_k\rangle\langle v_k| \ (x_k \geq 0). \tag{35}$$

Notice that any vector ξ on $\mathcal{H} \otimes \mathcal{H}$ is written by $\xi = \sum_i \xi_i \otimes \phi_i$ with some vectors $\xi_i \in \mathcal{H}$. Let $\sqrt{x_k} v_k = \sum_i \psi_{ki} \otimes \phi_i$, then $\sum_k x_k |v_k\rangle\langle v_k| = \sum_{i,j} (\sum_k |\psi_{ki}\rangle\langle\psi_{kj}|) \otimes E_{ij}$. Therefore, from (35), we obtain[*27]

$$\Phi(E_{ij}) = \sum_{k} |\psi_{ki}\rangle\langle\psi_{kj}| \ (\forall i, j = 1, \ldots, d). \tag{36}$$

Introduce the operators $V_k := \sum_i |\psi_{ki}\rangle\langle\phi_i|$ so that $V_k \phi_i = \psi_{ki}$. Then, by noting $\sum_i |\phi_i\rangle\langle\phi_i| = \mathbb{I}$, we have $\sum_k V_k A V_k^{\dagger} =$

[*26]Show that, indeed, any state ψ which has the maximal Schmidt number d can also be used for such a state.

[*27]Show that $\sum_{i,j=1}^{d} A_{ij} \otimes E_{ij} = 0 \Leftrightarrow A_{ij} = 0 \ (\forall i, j = 1, \ldots, d)$.

$\sum_k V_k(\sum_i |\phi_i\rangle\langle\phi_i|)A(\sum_j |\phi_j\rangle\langle\phi_j|)V_k^\dagger = \sum_{i,j,k}\langle\phi_i|A\phi_j\rangle(|V_k\phi_i\rangle\langle V_k\phi_j|) = \sum_{i,j}\langle\phi_i|A\phi_j\rangle(\sum_k |\psi_{ki}\rangle\langle\psi_{kj}|) = \sum_{i,j}\langle\phi_i|A\phi_j\rangle(\Phi(E_{ij}))$
$= \Phi(\sum_i |\phi_i\rangle\langle\phi_i|A\sum_j |\phi_j\rangle\langle\phi_j|) = \Phi(A)$ from (36) for any $A \in \mathcal{L}(\mathcal{H})$. Thus, the implication (iii)(iv) follows. Suppose (iv) holds. In order to show (i), it is enough to check the positivity of $\Phi \otimes id_n(|\psi\rangle\langle\psi|)$ for any $\psi \in \mathcal{H} \otimes \mathbb{C}^n$ and any $n \in \mathbb{N}$, since any positive operator on $\mathcal{H} \otimes \mathbb{C}^n$ is a positive linear combination of such forms (e.g., eigenvalue decomposition) and $\Phi \otimes id_n$ is linear. By expanding $\psi = \sum_{i=1}^d \phi_i \otimes \xi_i$, we observe $\Phi \otimes id_n(|\psi\rangle\langle\psi|) = \sum_{i,j}\Phi(E_{ij}) \otimes |\xi_i\rangle\langle\xi_j| = \sum_k \sum_{i,j} |V_k\phi_i\rangle\langle V_k\phi_j| \otimes |\xi_i\rangle\langle\xi_j| = \sum_k |x_k\rangle\langle x_k| \geq 0$ where $x_k := \sum_i (V_k\phi_i) \otimes \xi_i$. Thus, (i) follows, and this completes the whole proof. ■

Exercise 37. Let $T : \mathcal{L}(\mathcal{H}) \to \mathcal{L}(\mathcal{H})$ be a transposition map defined with a fixed orthonormal basis $\{\psi_i\}_{i=1}^d$ of $\mathcal{H}$ by $T(A) = \sum_{i,j}\langle\psi_j, A\psi_i\rangle|\psi_i\rangle\langle\psi_j|$. As a corollary of Theorem 4 (iii), show that T on $\mathcal{L}(\mathbb{C}^2)$ is positive, but not 2-positive.

This shows that m positivity does not necessary implies n positivity when $n > m$.

Exercise 38. Using the same technique in the proof of (iii)$\Rightarrow$ (iv) above, show that any Hermite map Φ on $\mathcal{L}(\mathcal{H})$ has a representation $\Phi(A) = \sum_i \epsilon_i V_i \rho V_i^\dagger$ where $\epsilon_i \in \{-1, 1\}$. Conversely, show that the map in this representation is Hermite.

The following is the representation theorem for trace-preserving completely positive map (hereafter TPCP map):

Theorem 5. *A linear map Φ on $\mathcal{L}(\mathcal{H})$ is TPCP map if and only if it has the representation*

$$\Phi(A) = \sum_k V_k A V_k^\dagger \ such\ that\ \sum_k V_k^\dagger V_k = \mathbb{I}. \qquad (37)$$

[Proof of Theorem 5] We have already seen that any linear map Φ is CP if and only if it has the form $\Phi(A) = \sum_k V_k A V_k^\dagger$ in Theorem 4. Let Φ be CP map with the form $\Phi(A) = \sum_k V_k A V_k^\dagger$. Then, for any $A \in \mathcal{L}(\mathcal{H})$, we have

$$\mathrm{tr}[\Phi(A)] = \mathrm{tr}[\sum_k V_k A V_k^\dagger] = \mathrm{tr}[A(\sum_k V_k^\dagger V_k)],$$

from the cyclic property of trace. This implies that the condition $\sum V_k^\dagger V_k = \mathbb{I}$ is necessary and sufficient for the CP map Φ to be trace preserving. ■

The operator sum representation (37) of TPCP map is known as **Kraus representation**, and operators V_k are called **Kraus operators**.

4. Open quantum mechanics

In the preceding section, we have explored a general time evolution map from the phenomenological point of view. A physically feasible dynamics should, however, be based on the law of quantum mechanics which is explained in Sec. 2. In postulate 5, the quantum dynamics has been only explained for an isolated quantum systems. A physical system is generally open, interacting with its environment, and thus a unitary time evolution cannot be applied directly to the system. If we include its environment so that the total system can be considered as an isolated system, a unitary time evolution can be applied to the total system. In this section, we explain the dynamics of open quantum mechanics using the mathematical tools given in the preceding section.

4.1. *General time evolution map for open quantum systems*

Let S be a quantum system described by d-dimensional Hilbert space $\mathcal{H}_S$, which is not isolated from the environment. To deal with the dynamics of S, one first has to take account of its environment E so that the total system $S + E$ can be (effectively) considered as an isolated system. Let $\mathcal{H}_E$ be the Hilbert space for the environment with d' dimension[*28]. Then, the total dynamics is described by a unitary time-evolution as $\rho_{tot}(t) = U_t \rho_{tot}(0) U_t^\dagger$ with a time evolution operator $U_t = \exp(-iHt)$ where H is a total Hamiltonian, an Hermitian operator on $\mathcal{H}_S \otimes \mathcal{H}_E$. Here, $\rho_{tot}(t)$ is a density operator on the total system $\mathcal{H}_S \otimes \mathcal{H}_E$ at time t.

However, since what we are interested in, or what we are restricted to observe, are observables on the system S only, the state should be described

---

[*28]Note that we have assumed that S is a finite quantum system described by a d-dimensional Hilbert space. However, one cannot generally assume that the dimension of the Hilbert space for environment is finite. Moreover, the environment may be even a quantum field with infinite degree of freedom, and in that case, even the description of the fixed Hilbert space might not be appropriate, requiring the theory of quantum fields (see for instance Ref. 35). However, this lecture note is intended to be logically closed by restricting to finite level systems, and therefore we assume the dimension of the environment is also finite. I believe this restriction is not essential to comprehend the whole structure of the theory of open quantum mechanics.

by the reduced density operator from $\rho_{tot}(t)$. Consequently, the dynamics for an open quantum system S is generally described by

$$\rho_s(t) = \mathrm{tr}_E(U_t\rho_{tot}(0)U_t^\dagger). \tag{38}$$

In particular, if the initial state is a product state $\rho_{tot}(0) = \rho_S \otimes \rho_E$, this time evolution can be considered as a map, from the initial reduced state ρ_S at $t = 0$ to the reduced state $\rho_S(t) = \mathrm{tr}_E(U_t\rho_S \otimes \rho_E U_t^\dagger)$ at time t. Consequently, the time evolution map*[29] $\Phi : \mathcal{S}(\mathcal{H}_S) \to \mathcal{S}(\mathcal{H}_S)$ is given by

$$\rho_S \mapsto \Phi(\rho_S) := \mathrm{tr}_E(U\rho_S \otimes \rho_E U^\dagger). \tag{39}$$

Exercise 39. Show that Φ is an affine map on $\mathcal{S}(\mathcal{H}_S)$.

From this and Exe. 30, Φ can be uniquely extended to the linear map on $\mathcal{L}(\mathcal{H}_S)$ which is

$$A \mapsto \Phi(A) := \mathrm{tr}_E(UA \otimes \rho_E U^\dagger) \;\forall A \in \mathcal{L}(\mathcal{H}_S). \tag{40}$$

This map is usually considered as the most general time-evolution map for the open quantum systems thanks to the following theorem:

Theorem 6. *A linear map Φ on $\mathcal{L}(\mathcal{H}_S)$ is given by (40) with a Hilbert space $\mathcal{H}_E$, a state $\rho_E \in \mathcal{S}(\mathcal{H}_E)$, and a unitary operator U on $\mathcal{H}_S \otimes \mathcal{H}_E$ if and only if Φ is a TPCP map on $\mathcal{L}(\mathcal{H}_S)$.*

[Proof of Theorem 6] Suppose that Φ is given by (40):

$$\Phi(A) := \mathrm{tr}_E[UA \otimes \rho_E U^\dagger] \; (A \in \mathcal{L}(\mathcal{H})).$$

---

*[29] It is worth noting the following: *when one considers a dynamics by means of the notion of a "map", one implicitly assumes that some quantities are fixed while applying the map to several initial condition (reduced state) of S.* Needless to say, the notion of maps has a domain — even though not necessarily including all the density operators on S — and is characterized by the way how it maps several elements in the domain. The fixing quantities in (39) are the environment, Hamiltonian (and hence U_t), and the state of the environment ρ_E (which is often taken to be thermal equilibrium state, in a real physical situation). Then, with the fixed terms $\mathcal{H}_E$, U, and ρ_E, one can test the dynamics to apply several initial states ρ_S of S with not only a mathematical point of view but also with an experimental point of view. The question of what kind of quantities should be fixed is a problem of physical setting. For instance, in some experiments, if the experimentalist has a strong belief that — even if he/she does not know the precise nature of environment — for each preparation of initial state of S, there exists the same environment and the interaction, and especially the same initial state of environment which is not correlated with the state of S, then the use of (39) will be quite useful to explain the phenomena for the experiments.

Let $\rho_E = \sum_i p_i |\phi_i\rangle\langle\phi_i|$ be an eigenvalue decomposition of ρ_E. Then, by introducing operators V_{ki} on $\mathcal{H}_S$ defined through $\langle\chi, V_{ki}\psi\rangle = \sqrt{p_i}\langle\chi \otimes \phi_k, U(\psi \otimes \phi_i)\rangle$ for any $\chi, \psi \in \mathcal{H}_S$, one can check

$$\Phi(A) = \sum_i p_i \operatorname{tr}_E[UA \otimes |\phi_i\rangle\langle\phi_i|U^\dagger]$$

$$= \sum_i p_i \sum_k \langle\phi_k|[UA \otimes |\phi_i\rangle\langle\phi_i|U^\dagger]|\phi_k\rangle = \sum_{k,i} V_{ki}AV_{ki}^\dagger, \qquad (41)$$

with the similar manner as in the final part of the proof of Theorem 4. Therefore, from Theorem 4 (i) and (iv) , Φ is CP. Moreover, we observe

$$\operatorname{tr}_S[\Phi(A)] = \operatorname{tr}_{SE}[UA \otimes \rho_E U^\dagger] = \operatorname{tr}_{SE}[A \otimes \rho_E U^\dagger U] = \operatorname{tr}_S[A]\operatorname{tr}_E[\rho_E] = \operatorname{tr}_S[A],$$

for any $A \in \mathcal{L}(\mathcal{H}_S)$. Therefore, Φ is TPCP. Conversely, suppose that Φ is TPCP, and therefore from Theorem 5, it has a Kraus representation (37). Let d' be the numbers of Kraus operators, and let $\mathcal{H}_E$ be a Hilbert space with dimension d' and let $\{e_k\}_{k=1}^{d'}$ be an orthonormal basis of $\mathcal{H}_E$. From Exe. 26, there exists a unitary operator U on $\mathcal{H}_S \otimes \mathcal{H}_E$ such that $U(\psi \otimes e_1) = \sum_{k=1}^{d'}(V_k\psi) \otimes e_k$. Let $\rho_E := |e_1\rangle\langle e_1|$ be a pure state on $\mathcal{H}_E$. Then, for any $A \in \mathcal{L}(\mathcal{H}_S)$, it is easy to see

$$\operatorname{tr}_E[UA \otimes \rho_E U^\dagger] = \sum_k V_k AV_k^\dagger = \Phi(A).$$

Therefore, any TPCP map can be written in the form (40). ∎

This fact is important for a general treatment of the open quantum dynamics. Indeed, we have shown that with plausible assumptions, the general time evolution map should be TPCP. On the other hand, Theorem 6 implies that any TPCP map can be physically realizable as an open quantum dynamics based on the postulates of quantum mechanics. Consequently, one can say that the most general time-evolution map for open quantum systems is a trace-preserving completely positive map[*30], while the most general time-evolution map is a unitary map for isolated quantum systems.

[*30]Since we have restricted to the case of no-initial correlations in (39), one might think that there are another kinds of open quantum dynamics which can not be explained by TPCP map, if one consider more general situation with initial correlations (38). Indeed, this is in some sense true and one can say the most general dynamics in open quantum systems is described by (38). The problem occurs when one would like to describe it by means of a time evolution map. Indeed, there are even several attempts to consider a time evolution map with initial correlations by means of non TPCP map. However, in dealing with this topics, I strongly recommend to read the footnote *29 about a time-evolution map. Indeed, one can mathematically imagine various situations to use a time evolution map to explain the case of initial correlations, however sometimes there are meaningless conflicts by raising mutually contradictory results; one insists that time evolution map

4.2. *Time evolution with non-decreasing entropy*

Now, let us see some general property of the open quantum dynamics described by TPCP maps. One important class of time-evolution maps is a class of maps which does not decrease entropy. For the entropy, let us consider the Tsallis entropy, among which the most important one is the von Neumann entropy. In the following, we show that such a class of time evolution maps is given by the set of all the **unital** TPCP maps: Here, a linear map Φ on $\mathcal{H}(\mathcal{H})$ is said to be unital if $\Phi(\mathbb{I}) = \mathbb{I}$:

Exercise 41. Show that CP map Φ is unital if and only if it can be written as

$$\Phi(A) = \sum_k V_k A V_k^\dagger \text{ such that } \sum_k V_k V_k^\dagger = \mathbb{I}. \qquad (42)$$

Exercise 42. Show that the sets of trace-preserving maps, positive maps, n-positive maps, completely positive maps, unital maps, and the combination of these maps, are all convex sets.

Physically, if a time-evolution map is described by a unital TPCP map Φ, this means that the maximal mixed state $\frac{\mathbb{I}}{d}$ is a fixed point of Φ. Since $\frac{\mathbb{I}}{d}$ is the unique state which minimize the Tsallis entropy (and thus von Neumann entropy), this implies that a time evolution map Φ described by TPCP map is unital if Φ does not decrease an entropy. Remarkably, the opposite is also true, and thus we have

Theorem 7. *A time evolution map Φ described by TPCP map does not decrease an entropy if and only if it is unital.*

---

with initial correlations is not TPCP, while another insists that it is indeed TPCP even with initial correlations. Most of such contradictions seem not to be mathematical one, but due to the different physical settings while it is difficult to find such considerations in many references. One should rethink the use of time evolution map as in the footnote *29. Nevertheless, one might be interesting to reconsider what kind of assumptions which have leaded us to be TPCP map for a general description of time evolution map in the preceding sections can break up in the case of initial correlations:

Exercise 40. Consider the above problems. [Hint: For instance, the implicit assumption that the state preparation is independent of the dynamics is too strong. In the case of initial correlations, the initial reduced state $\rho_S(0) := \mathrm{tr}_E \rho_{tot}(0)$ has a particular form such that it comes from the total state $\rho_{tot}(0)$, and one can not apply another preparation even with the same density operator; Also from the reflection of the footnote *29, consider what kind of quantities should be fixed in time evolution map?]

To prove this, first we will show the following lemma:

Lemma 1. *Let A be an Hermitian operator on $\mathcal{H}$, and let Φ be a unital TPCP map on $\mathcal{L}(\mathcal{H})$. Let $A = \sum_{i=1}^{d} a_i |\psi_i\rangle\langle\psi_i|$ and $\Phi(A) = \sum_{i=1}^{d} a_i' |\psi_i'\rangle\langle\psi_i'|$ be eigenvalue decompositions of A and $\Phi(A)$, respectively. Then, there exists a $d \times d$ doubly stochastic matrix B such that $a_i' = \sum_{i=1}^{d} B_{ij} a_j$.*

A $d \times d$ matrix B is called a stochastic matrix if $B_{ij} \geq 0$ for all $i, j = 1, \ldots, d$ and $\sum_i B_{ij} = 1$ for all $j = 1, \ldots, d$. It is called a doubly stochastic matrix if B is a stochastic matrix and $\sum_{j=1} B_{ij} = 1$ for all $i = 1, \ldots, d$.

Proof of Lemma 1 We observe

$$a_i' = \langle\psi_i', \Phi(A)\psi_i'\rangle = \sum_j \langle\psi_i', \Phi(|\psi_j\rangle\langle\psi_j|)\psi_i'\rangle a_j =: \sum_j B_{ij} a_j,$$

where $B_{ij} := \langle\psi_i', \Phi(|\psi_j\rangle\langle\psi_j|)\psi_i'\rangle$. It is easy to see $B_{ij} \geq 0$, and $\sum_i B_{ij} = \sum_i B_{ij} = 1$ from the property of unital CPTP map of Φ. ∎

In order to prove Theorem 7, Birkhoff-von Neumann theorem[36] can be used:

Theorem 8. *A $d \times d$ matrix is doubly stochastic if and only if it is a convex combination of permutation matrices.*

A $d \times d$ matrix P is called a permutation matrix if it is obtained by permuting the rows of an identity matrix according to some permutation of the numbers 1 to d (Show a permutation matrix is a doubly stochastic matrix).

Exercise 43. Prove Birkhoff-von Neumann theorem.

Proof of Theorem 7 [If part] Let Φ be a unital TPCP map on $\mathcal{L}(\mathcal{H})$. In the following, we will see that $S_q(\rho) \leq S_q(\Phi(\rho))$ for any $\rho \in \mathcal{S}(\mathcal{H})$. Let $\{a_j\}, \{a_j'\}$ be the set of eigenvalues of ρ and $\Phi(\rho)$. From definition (16), we have $S_q(\rho) = H_q(\{a_i\})$, $S_q(\Phi(\rho)) = H_q(\{a_i'\})$. Then, from Lemma 1, we have $a_i' = \sum_j B_{ij} a_j$ with a doubly stochastic matrix B. By the Birkhoff-von Nenmann theorem, we have $B_{ij} = \sum_k p_k P_{ij}^{(k)}$ where p_k is a probability distribution and $P^{(k)}$ are permutation matrices. Therefore, we have

$$a_i' = \sum_j B_{ij} a_j = \sum_k p_k a_i^{(k)},$$

where $a_i^{(k)} = \sum_j P_{ij}^{(k)} a_j$. Note that Tsallis entropy H_q is a permutation invariant function, and thus we have $H_q(\{a_i^{(k)}\}) = H(\{a_i\})$. Finally, using the concavity of H_q, we obtain

$$S_q(\Phi(\rho)) = H_q(\{a_i'\}) \geq \sum_k p_k H_q(\{a_i^{(k)}\}) = \sum_k p_k H_q(\{a_i\}) = H_q(\{a_i\}) = S_q(\rho).$$

36

Thus, we have proved the if part.

[Only if part] Let Φ be a TPCP map which does not decrease a Tsallis entropy. Since the maximal mixed state $\frac{1}{d}\mathbb{I}$ is the unique state which maximize Tsallis entropy, Φ should map $\frac{1}{d}\mathbb{I}$ to $\frac{1}{d}\mathbb{I}$, otherwise it contradicts the non-decreasing property of entropy. Therefore, this implies Φ is unital.

∎

As a corollary, we also have

Theorem 9. *A time evolution map Φ described by TPCP map does not increase the purity if and only if it is unital.*

Proof Apply Theorem 7 for the Linear entropy ($q = 2$). ∎

Theorem 7 also leads an operational meaning of quantum Tsallis entropy with respect to the classical one:

Theorem 10. *Quantum Tsallis entropy is the minimal classical Tsallis entropy among non-degenerate observables.*

Exercise 44. Let $A = \sum_i a_i |\phi_i\rangle\langle\phi_i|$ be an arbitrary non-degenerate observables, and let $p^A(\rho)_i = \langle\phi_i, \rho\phi_i\rangle$ be the probability to obtain a_i from the measurement of A under state ρ. Then, prove that the map $\Phi(A) := \sum_i p^A(\rho)_i |\phi_i\rangle\langle\phi_i|$ is a unital TPCP map.

Proof of Theorem 10 Under the same notion as Exe. 44, we observe

$$S_q(\Phi(\rho)) = S_q(\sum_i p^A(\rho)_i |\phi_i\rangle\langle\phi_i|) = H_q(p^A(\rho)_i),$$

since $\{p^A(\rho)_i\}$ is a set of eigenvalues of $\Phi(\rho)$. From Theorem 7, we have

$$H_q(p^A(\rho)_i) \geq S_q(\rho).$$

Therefore, any Tsallis entropy among non-degenerate observables is bounded by quantum Tsallis entropy from below. Let $\rho = \sum_i p_i |\rho_i\rangle\langle\rho_i|$ be an eigenvalue decomposition of ρ. By considering a non-degenerate observables B with eigenvectors $\{\rho_i\}$, one obtains that

$$H_q(p^B(\rho)_i) = S_q(\rho),$$

since $p^B(\rho)_i = p_i$. ∎

4.3. *Quantum Markovian Dynamics*

So far, we have seen an open quantum dynamics as a discrete map, forgetting the time. In the following, we treat the dynamics following a continuous time dependence, and we focus on the time differential equation of the dynamics, which is called a **master equation**. The von Neumann equation (24) is a typical master equation. However, it is a dynamics only for an isolated system, and what we are going to deal with is a dynamics for open quantum systems with decoherence.

First, let us see the physical basis of master equation from the microscopic point of view. Let H_{tot} be a total Hamiltonian of the total system $S + E$, with which the time-evolution map is given by the unitary operator $U_t = \exp(-iH_{tot}t)$. Physically, it is often written as

$$H_{tot} = H_S \otimes \mathbb{I}_E + H_{int} + \mathbb{I}_S \otimes H_E,$$

where H_S, H_E are the free Hamiltonians for S and E, while H_{int} is an interaction Hamiltonian. If $H_{int} = 0$, then $U_t = \exp(-iH_St) \otimes \exp(-iH_Et)$, and one can show the reduced dynamics is unitary with respect to $\exp(-iH_St)$. From this fact, it is an interaction Hamiltonian that causes decoherence, a discord from the unitary dynamics. Moreover, the mechanism of the process to increase an entropy, or to decrease the purity for a system S can be understood as follows: First, let the system S is in a pure state. Then, according to Theorem 2, there are no correlations with environment at the initial time. However, since there is an interaction term H_{int} which generally causes correlations as time goes by, the reduced state comes to be a mixed state, again from the result of Theorem 2.

The von Neumann-Schrödinger equation (24) for the total system is

$$\frac{d}{dt}\rho_{tot}(t) = -i[H_{tot}, \rho_{tot}(t)].$$

By taking a partial trace over the environment, one obtains the master equation for the system S:

$$\frac{d}{dt}\rho_S(t) = -i[H_S, \rho_S(t)] - i\,\mathrm{tr}_E([H_{int}, \rho_{tot}(t)]), \tag{43}$$

where we have used (i) the commutativity of $\frac{d}{dt}$ and the partial trace operation, (ii) the commutativity of the operation $-i[H_S, \cdot]$ and the partial trace operation, and (iii) the equality $\mathrm{tr}_E(H_E\rho_{tot}(t)) = \mathrm{tr}_E(\rho_{tot}(t)H_E)$.

Exercise 45. Explain these relations in details.

One observes that there is an additional term in the right hand side of equation (43) compared with the von Neumann-Schrödinger equation (24). Therefore one can consider this term the cause for the decoherence. To calculate this term with given Hamiltonians is generally quite a difficult problem; we usually need to have some techniques of approximations.[8,9,11,12]

In particular, we restrict to a class of **Markovian dynamics**,[7] where one can predict all the future time only from the information at time t, and it has been proved to be useful in broad range of quantum physics.

More precisely, let $\Phi_t : \mathcal{S}(\mathcal{H}) \to \mathcal{S}(\mathcal{H})$ be a time-evolution map from initial density operator to a density operator at time t, especially defined for a future $t \geq 0$. Markovian property is defined to have a semigroup law $\Phi_{t+s} = \Phi_t \Phi_s$. As we have seen, the map should have a completely positive map in order to have a physical promise. Moreover, it is natural to assume a continuity of a time $t \mapsto \Phi_t(\rho)$ for any state ρ, here with respect to the trace norm: $||\Phi_{t+s}(\rho) - \Phi_t(\rho)||_1 \to 0$ $(as\ s \to 0)$. These define a **completely positive dynamical semigroup** (hereafter CPDS),[39-41] which provides a general framework of a quantum Markovian process. Mathematically, a one parameter family $\{\Phi_t \mid t \geq 0\}$ of a CPDS is a strongly continuous and contractive one parameter semigroup on the Banach space of $\mathcal{L}(\mathcal{H})_h$ with the trace norm.

Exercise 46. Prove the contraction of Φ_t is automatically satisfied by the trace preserving property of Φ_t, i.e., prove that $||\Phi_t|| := \sup_{A \in \mathcal{L}(\mathcal{H}), ||A||_1 = 1} ||\Phi_t(A)||_1 \leq 1$.

Exercise 47. Check that a unitary map $\mathcal{U}_t$ in (25) is a CPDS.

The generator of $\{\Phi_t\}$ is defined by

$$\mathcal{L}\,A = \lim_{s \to 0} \frac{\Phi_s A - A}{s},$$

where the Domain of A is the subset of $\mathcal{L}(\mathcal{H})_h$ where the limit exists. From the Hille-Yoshida theory of semigroup,[38] it is guaranteed that $\mathcal{L}$ is a densely defined closed operator, and we have the differential equation

$$\frac{d}{dt}\Phi_t(A) = \mathcal{L}\,\Phi_t(A) = \Phi_t(\mathcal{L}\,A),$$

for any A in the domain of $\mathcal{L}$. Moreover, in the finite level systems, where the dimension $\mathcal{L}(\mathcal{H})_h$ is finite, $\mathcal{L}$ is bounded operator on $\mathcal{L}(\mathcal{H})_h$, and we have

$$\Phi_t = \exp(t\,\mathcal{L}) := \sum_{n=1}^{\infty} \frac{(t\,\mathcal{L})^n}{n!}.$$

Gorini, Kossakowski, Sudarshan[41] obtained a general representation form of a generator of a completely positive dynamical semigroup for finite level systems:

Theorem 11. *A linear map $\mathcal{L} : \mathcal{L}(\mathcal{H}) \to \mathcal{L}(\mathcal{H})$ is the generator of a CPDS $\{\Phi_t\}$ if and only if it can be expressed in the form*

$$\mathcal{L}(\rho) = -i[H, \rho] + \frac{1}{2} \sum_{i,j=1}^{d^2-1} c_{ij}([F_i, \rho F_j^\dagger] + [F_i \rho, F_j^\dagger]), \qquad (44)$$

where $H \in \mathcal{O}(\mathcal{H})$, $\operatorname{tr} F_i^\dagger F_j = \delta_{ij}$ $(i, j = 1, \ldots, d^2)$ where $F_{d^2} := \frac{1}{\sqrt{d}} \mathbb{I}$, and $\{c_{ij}\}$ is a complex $(d^2 - 1) \times (d^2 - 1)$ positive matrix. For a given $\mathcal{L}$, H is uniquely determined by the condition $\operatorname{tr} H = 0$ and c_{ij} is uniquely determined by the choice of the F_i $(i = 1, \ldots, d^2 - 1)$.

Independently, Lindblad obtained the same representation theorem in infinite dimensional cases for bounded generators.[40] Therefore, we call the master equation $\frac{d}{dt}\rho(t) = \mathcal{L}(\rho_t)$ with the generator (44) GKLS (Gorini-Kossakowski-Lindblad-Sudarshan) master equation.

The following is a typical heuristic proof of the "only if part" of Theorem 11: Let Φ_t be CPTP map, and then the generator is defined by $\mathcal{L}(A) := \lim_{s \to 0} \frac{\Phi_s(A) - A}{s}$. Let $\Phi_t(A) = \sum_k V_k(t) A V_k^\dagger(t)$ be a Kraus Representation of Φ_t at time t. Since $\{F_i\}$ forms an orthonormal basis of $\mathcal{L}(\mathcal{H})$, the Kraus operators can be written as $V_k(t) = \sum_{i=1}^{d^2} a_{ki}(t) F_i$. Then, we have

$$\Phi_t(A) = \sum_{ij} b_{ij}(t) F_i A F_j^\dagger,$$

where $b_{ij}(t) := \sum_k a_{ki}(t) \overline{a_{kj}(t)}$ which is a positive matrix from the form. We observe

$$\frac{\Phi_s(A) - A}{s} = \left(\frac{1}{s} \frac{b_{d^2 d^2}(s)}{d} - 1 \right) A + \left(\sum_{i=1}^{d^2-1} \frac{1}{\sqrt{d}s} b_{id^2}(s) F_i \right) A$$
$$+ A \left(\sum_{j=1}^{d^2-1} \frac{1}{\sqrt{d}s} b_{d^2 j}(s) F_j^\dagger \right) + \sum_{i,j=1}^{d^2-1} \frac{b_{ij}(s)}{s} F_i A F_j^\dagger. \quad (45)$$

Let

$$c_{d^2 d^2} := \lim_{s \to 0} \left(\frac{1}{s} \frac{b_{d^2 d^2}(s)}{d} - 1 \right), \qquad (46a)$$

$$F := \lim_{s \to 0} \left(\sum_{j=1}^{d^2-1} \frac{1}{\sqrt{d}s} b_{d^2 j}(s) F_j^\dagger \right) \qquad (46b)$$

$$c_{ij} := \lim_{s \to 0} \frac{b_{ij}(s)}{s}. \qquad (46c)$$

Then we obtain

$$\mathcal{L}(A) = c_{d^2 d^2} A + F^\dagger A + AF + \sum_{i,j=1}^{d^2-1} c_{ij} F_i A F_j^\dagger.$$

By decomposing $F = F_R + iF_I$ ($F_R = \frac{F+F^\dagger}{2}$, $F_I = \frac{F-F^\dagger}{2i}$), it follows

$$\mathcal{L}(A) = -i[F_I, A] + [F_R + \frac{c_{d^2 d^2}}{2}, A]_+ + \sum_{i,j=1}^{d^2-1} c_{ij} F_i A F_j^\dagger.$$

Furthermore, let $H = F_I$ and $G = F_R + \frac{c_{d^2 d^2}}{2}$, and rewrite the equation as

$$\mathcal{L}(A) = -i[H, A] + [G, A]_+ + \sum_{i,j=1}^{d^2-1} c_{ij} F_i A F_j^\dagger.$$

Finally, using $\operatorname{tr} L(A) = 0$ for all $A \in \mathcal{L}(\mathcal{H})$ from the trace preserving property of Φ_t, we obtain

$$\operatorname{tr}(2G + \sum_{i,j=1}^{d^2-1} c_{ij} F_j^\dagger F_i) A \ \Rightarrow\ G = -\frac{1}{2} \sum_{i,j=1}^{d^2-1} c_{ij} F_j^\dagger F_i.$$

This completes the derivation of GKSL equation (44).

Exercise 48. The above "proof" is not yet rigor. By using the existence of limit $\mathcal{L}(A) := \lim_{s \to 0} \frac{\Phi_s(A) - A}{s}$, prove the existence of limits in (46). This completes the proof of only if part of the theorem.

Exercise 49. Prove the if part of Theorem 11.

Acknowledgment The author would like to appreciate all the staffs of the summer school, " Symposium on Decoherence Suppression in Quantum Systems", Kinki University Quantum Computing Series, September 7-10 2008 at Oxford Kobe Institute (Kobe, Japan), especially Prof. Mikio Nakahara and Dr. Yukihiro Ota for their hospitality and useful discussion. My special thanks is to Prof. Kossakowski, my supervisor during my visit to Torun, Poland, who is one of the founders of (completely positive) dynamical semigroup.

Appendix A. Mathematical Background

The set of complex numbers, reals numbers, and natural numbers are denoted by $\mathbb{C}$, $\mathbb{R}$, and $\mathbb{N}$, respectively. A complex conjugate of $z \in \mathbb{C}$ is denoted by $\bar{z}$.

[Direct product] Let S and T be sets. The **direct product set**, denoted by $S \times V$, is the set of all the pairs (s, v) of the elements $(s \in S, v \in V)$.

[Metric space] A set S is called a **metric space** if it has a distance operation $d : S \times S \to \mathbb{R}$ satisfying (i) $d(s_1, s_2) \geq 0$ for any $s_1, s_2 \in S$, (ii) $d(s_1, s_2) = 0$ iff $s_1 = s_2$, (iii) $d(s_1, s_2) = d(s_2, s_1)$, (iv) for $s_1, s_2, s_3 \in S$, $d(s_1, s_3) \leq d(s_1, s_2) + d(s_2, s_3)$ (triangle inequality). A sequence $s_n \in S$ $(n = 1, 2, \ldots)$ in a metric space is called a **Cauchy sequence** if for any $\epsilon > 0$ there exists $n_0 \in \mathbb{N}$ such that $d(s_n, s_m) < \epsilon$ for any $n, m \geq n_0$. A sequence $s_n \in S$ $(n = 1, 2, \ldots)$ is said to be a **convergent sequence** if there exists $s \in S$ such that for any $\epsilon > 0$ there exists $n_0 \in \mathbb{N}$ such that $d(s, s_n) < \epsilon$ for any $n \geq n_0$.

Exercise 50. Show that any convergent sequence is a Cauchy sequence.

[Completeness] A metric space S is said to be **complete** if any Cauchy sequence is a convergent sequence.

[Linear space] A set V is called a complex (resp. real) **linear space**, or **vector space**, on a complex (resp. real) number field $\mathbb{C}$ (resp. $\mathbb{R}$) if there exists a sum $v_1 + v_2 \in V$ for every pair of v_1 and v_2 of vectors in V and a scalar product αv for $\alpha \in K$ (hereafter $K = \mathbb{R}$ or $\mathbb{C}$) and $v \in V$ in a way that (S-i) $v_1 + v_2 = v_2 + v_1$, (S-ii) $v_1 + (v_2 + v_3) = (v_1 + v_2) + v_3$, (S-iii) there exists a vector 0 such that $v + 0 = v$ for any $v \in V$, (S-iv) for any $v \in V$ there exists a vector $-v$ such that $v + (-v) = 0$; (P-i) $\alpha(\beta v) = (\alpha \beta)v$, (P-ii) $1v = v$ for any $v \in V$, (SP-i) $\alpha(v_1 + v_2) = \alpha v_1 + \alpha v_2$, (SP-ii) $(\alpha + \beta)v = \alpha v + \beta v$.

[Subspace] A non-empty subset W of V is called a **subspace** if $\alpha w_1 + \beta w_2 \in W$ for any $w_1, w_2 \in W$ and $\alpha, \beta \in K$.

Let V be a linear space on K.

[Normed space] V is called a **normed space** if it has a **norm** operation $V \to \mathbb{R}$, denoted by $v \in V \mapsto ||v|| \in \mathbb{R}$ which satisfies (i) $||v|| \geq 0$ for any $v \in V$, (ii) for $v \in V$, $||v|| = 0 \Leftrightarrow v = 0$, (iii) for $v \in V, \alpha \in K$, $||\alpha v|| = |\alpha| ||v||$, and (iv) for $v_1, v_2 \in V$, $||v_1 + v_2|| \leq ||v_1|| + ||v_2||$ (triangle inequality).

Exercise 51. Show that $d(v_1, v_2) := ||v_1 - v_2||$ is a distance, and thus a normed space is a metric space.

[Banach space] A normed space V is called a **Banach space** if V is complete with respect to the metric induced by the norm.

[Inner product space] V is called an **inner product space** if it has an **inner product** operation from $V \times V$ to K, written as $(v_1, v_2) \in V \times V \mapsto$

$\langle v_1, v_2 \rangle \in K$ which satisfies (i) $\langle v, v \rangle \geq 0$ for any $v \in V$, (ii) for $v \in V$, $\langle v, v \rangle = 0 \Leftrightarrow v = 0$, (ii) for $v_1, v_2 \in V$, $\overline{\langle v_1, v_2 \rangle} = \langle v_2, v_1 \rangle$, and (iv) for $v, w_1, w_2 \in V, \alpha, \beta \in K$, $\langle v, \alpha w_1 + \beta w_2 \rangle = \alpha \langle v, w_1 \rangle + \beta \langle v, w_2 \rangle$.

Exercise 52. Prove that the $|\langle v, w \rangle|^2 \leq |\langle v, v \rangle| |\langle w, w \rangle|$ for any $v, w \in V$ (Schwartz inequality).

Exercise 53. Show that the $||v|| := \sqrt{\langle v, v \rangle}$ is a norm, and thus an inner product space is a normed space.

We say $v_1 \in V$ and $v_2 \in V$ are **orthogonal** if $\langle v_1, v_2 \rangle = 0$, and $v \in V$ is called a **unit vector** if it has a unit norm $||v|| = 1$.
[Hilbert space] An inner product space V is called a **Hilbert space** if V is complete with respect to the metric induced by the inner product.

Exercise 54. Show that $\mathbb{C}^d := \mathbb{C} \times \mathbb{C} \times \cdots \times \mathbb{C}$ is a d dimensional complex Hilbert space for $c = (c_1, c_2, \ldots, c_d), c' = (c'_1, c'_2, \ldots, c'_d) \in \mathbb{C}^d$, $\alpha, \beta \in \mathbb{C}$, $\alpha(c_1, c_2, \ldots, c_d) + \beta(c'_1, c'_2, \ldots, c'_d) := (\alpha c_1 + \beta c'_1, \alpha c_2 + \beta c'_2, \ldots, \alpha c_d + \beta c'_d)$ with the inner product $\langle c, c' \rangle := \sum_{i=1}^d \bar{c}_i c'_i$. Similarly, define $\mathbb{R}^d$ and show it is a real d dimensional Hilbert space.

To deal with vector together with matrix, it is convenient to write a vector in $\mathbb{C}^d$ as a column vector:

$$\begin{pmatrix} c_1 \\ c_2 \\ \vdots \\ c_d \end{pmatrix} =: (c_1, \ldots, c_d)^T,$$

where T denotes the transposition operation.

Exercise 55. Verify that any normed space (resp. inner product space) with finite dimensions is a **Banach space** (resp. **Hilbert space**).

Consequently in this note, *one can consider Hilbert space just an inner product space*, since we only deal with finite level systems. Hereafter, let $\mathcal{H}$ be a d dimensional Hilbert space.

Exercise 56. First by showing that a subspace W of $\mathcal{H}$ is closed, verify that for any subspace W and for any $\psi \in \mathcal{H}$, there exists the unique vector ϕ in W which is closest to ψ, i.e., for any $\phi' \in W$, it satisfies $||\psi - \phi|| \leq ||\psi - \phi'||$.

[Orthonormal system, and orthonormal basis] A set of vectors $\{\phi_i\}_{i=1}^m$ ($m \in \mathbb{N}$) of $\mathcal{H}$ is called an **orthonormal system** if $\langle \phi_i, \phi_j \rangle = \delta_{ij}$ ($\forall i, j =$

$1, \ldots, m$). If $m = d$, it is called **orthonormal basis** (or complete orthonormal system). (Show that an orthonormal system $\{\phi_i\}_{i=1}^d$ of $\mathcal{H}$ is a linearly independent set.)

Exercise 57. Show that for any $\psi \in \mathcal{H}$,

$$\psi = \sum_{i=1}^d \langle \phi_i, \psi \rangle \phi_i. \tag{A.1}$$

Exercise 58. Show that any d dimensional complex Hilbert space $\mathcal{H}$ is homomorphic to $\mathbb{C}^d$, i.e., there exists a bijective function $f : \mathcal{H} \to \mathbb{C}^d$ s.t. (i) $f(\alpha\psi + \beta\phi) = \alpha f(\psi) + \beta f(\phi)$ for all $\alpha, \beta \in \mathbb{C}, \psi, \phi \in \mathcal{H}$; (ii) $\langle \psi, \phi \rangle = \langle f(\psi), f(\phi) \rangle$. (Hint: use f such that $f(\psi) = (\langle \phi_1, \psi \rangle, \ldots, \langle \phi_d, \psi \rangle) \in \mathbb{C}^d$ for some fixed orthonormal basis $\{\phi_i\}_{i=1}^d$.)

Therefore, one can always identify a d-dimensional Hilbert space $\mathcal{H}$ with $\mathbb{C}^d$.

[Orthogonal complement] **Orthogonal complement** of a subspace W is the subspace $W^\perp$ defined by $W^\perp := \{\phi \in \mathcal{H} \mid \langle \psi, \phi \rangle = 0 \ (\forall \psi \in \mathcal{H})\}$.

Exercise 59. (**Projection theorem**) Let W be a subspace of $\mathcal{H}$. Show that every $\psi \in \mathcal{H}$ can be uniquely decomposed as $\psi = \phi + \xi$ with $\phi \in W$ and $\xi \in W^\perp$. (Hint: ϕ is indeed the closed vector in W to ψ of which the existence has been shown in Exercise 56.)

[Linear operators] A map $A : \mathcal{H} \to \mathcal{H}$ is called a **linear operator** if $A(\alpha v_1 + \beta v_2) = \alpha A(v_1) + \beta A(v_2)$. Hereafter, an operator always means a linear operator. The range of A, denoted as ran$A$, is the set ran$A := \{\psi \in \mathcal{H} \mid \exists \phi \in \mathcal{H}, \ s.t. \ \psi = A\phi\}$. (Show that ran A is a subspace of $\mathcal{H}$.)
[Examples of operators] An identity operator and zero operator, denoted as $\mathbb{I}$ and 0, are linear operators defined by $\psi \in \mathcal{H}$, $\mathbb{I}\psi = \psi, 0\psi = 0$. For vectors $\psi, \phi \in \mathcal{H}$, the operator $|\psi\rangle\langle\phi|$ is defined by

$$|\psi\rangle\langle\phi| \ \xi := \langle \phi, \xi \rangle \psi$$

for all $\xi \in \mathcal{H}$. (Prove that these are linear operators.)

Let $\mathcal{L}(\mathcal{H})$ be the set of linear operators on $\mathcal{H}$. One can naturally introduce (a) the operator sum and (b) the operator scalar product: (a) For $A, B \in \mathcal{L}(\mathcal{H})$, $A+B \in \mathcal{L}(\mathcal{H})$ is an operator defined by $(A+B)\psi := A\psi + B\psi$; (b) for $A \in \mathcal{L}(\mathcal{H}), \alpha \in \mathbb{C}$, αA is an operator defined by $(\alpha A)\psi := \alpha(A\psi)$. Then it is easy to see that $\mathcal{L}(\mathcal{H})$ is a complex vector space. Moreover, one can introduce (c) the operator multiplication $AB \in \mathcal{L}(\mathcal{H})$ defined by

$(AB)\psi := A(B\psi)$, whereby $\mathcal{L}(\mathcal{H})$ forms an algebra. Notice that in general $AB \neq BA$, i.e., they are not commutative.

[Eigenvalues] For $A \in \mathcal{L}(\mathcal{H})$, a complex number $\lambda \in \mathbb{C}$ is called an eigenvalue of A if there exists a nonzero vector $\psi \in \mathcal{H}$ such that $A\psi = \lambda\psi$. The vector ψ is called an eigenvector of A corresponding to the eigenvalue λ. A subset $E_A(\lambda) \subset \mathcal{H}$ defined by $E_A(\lambda) = \{\psi \in \mathcal{H} \mid A\psi = \lambda\psi\}$ is called an eigenspace of A with an eigenvalue λ.

Exercise 60. Prove that $E_A(\lambda)$ is a subspace of $\mathcal{H}$, and $\lambda \in \mathbb{C}$ is an eigenvalue of A if and only if dim $E_A(\lambda) \geq 1$. (If dim $E_A(\lambda) = 0$, then $E_A(\lambda) = \{0\}$.)

The dimension of $E_A(\lambda)$ is called a degeneracy of A for eigenvalue λ. If a degeneracy of A for λ is greater or equal to 2, then λ is called a degenerate eigenvalue of A.

Exercise 61. Show that an operator A of $\mathcal{H}$ can be defined by the matrix elements $\langle \psi, A\phi \rangle$ for any $\psi, \phi \in \mathcal{H}$. By showing the operator identity:

$$\langle \phi, A\psi \rangle = \frac{1}{4}\Big(\langle \phi + \psi, A\phi + \psi \rangle - \langle \phi - \psi, A\phi - \psi \rangle$$
$$+ i\langle \phi + i\psi, A\phi + i\psi \rangle - i\langle \phi - i\psi, A\phi - i\psi \rangle\Big), \qquad (A.2)$$

show that it is indeed enough to specify the matrix elements $\langle \psi, A\psi \rangle$ for any $\psi \in \mathcal{H}$ to define operator A.

[Adjoint of operator] **Adjoint operator** $A^\dagger$ of $A \in \mathcal{L}(\mathcal{H})$ is defined through

$$\langle \psi, A^\dagger \phi \rangle = \langle A\phi, \psi \rangle$$

for any $\psi, \phi \in \mathcal{H}$.

Exercise 62. Show that $|\psi\rangle\langle\phi|^\dagger = |\phi\rangle\langle\psi|$, $(\phi, \psi \in \mathcal{H})$.

Exercise 63. Prove that $\dagger$ operation is an involution on $\mathcal{L}(\mathcal{H})$: i.e., (i) $(A^\dagger)^\dagger = A$, (ii) $(A+B)^\dagger = A^\dagger + B^\dagger$, (iii) $(\alpha A)^\dagger = \bar{\alpha}A^\dagger$, (iv) $(AB)^\dagger = B^\dagger A^\dagger$, (v) $\|A^\dagger\|_\infty = \|A\|_\infty$ (for (v), see the operator norm below).

[Commutator] The commutator $[A, B]$ for $A, B \in \mathcal{L}(\mathcal{H})$ is defined by $[A, B] := AB - BA$.

[Hermitian, Unitary, Positive, and Projection Operators] Operator A is called (a) a **normal operator** if $[A, A^\dagger] = 0$, (b) an **Hermitian operator** (or a **self-adjoint operator**) if $A = A^\dagger$, (c) a **unitary operator** if

$AA^\dagger = A^\dagger A = \mathbb{I}$, (d) a **positive operator** (or positive semidefinite operator), denoted as $A \geq 0$, if $\langle \psi, A\psi \rangle \geq 0$ for all $\psi \in \mathcal{H}$, and (e) a **projection operator** (or simply projection) if $A = A^\dagger = A^2$. Two positive operators A and B are said to be orthogonal if $AB = 0$.

Exercise 64. Prove that eigenvalues of Hermitian operators and positive operators are real and positive, respectively. Prove also that eigenvalues of projection operators are 0 or 1.

Exercise 65. Prove that A is an Hermitian operator if and only if $\langle \psi, A\psi \rangle \in \mathbb{R}$ for all $\psi \in \mathcal{H}$. (Hint: use the operator identity (A.2).)

Exercise 66. Prove that (i) if A is a projection operator, A is positive; (ii) if A is positive, A is Hermitian; (iii) if A is either unitary, or Hermitian, A is normal. (Hint: (ii) use Exercise 65.)

Exercise 67. Prove that $A \geq 0$ if and only if A is Hermitian and all the eigenvalues are positive. [Hint: for only if part, use Exercise 65; for if part use the spectral decomposition (see Exercise 75 below).]

Exercise 68. Prove that for any unit vector $\psi \in \mathcal{H}$, $|\psi\rangle\langle\psi|$ is a projection operator. It is the one dimensional projection operator onto the subspace $\{\alpha\psi \mid \alpha \in \mathbb{C}\}$ (See Exercise 71 below).

Exercise 69. Prove that $A^\dagger A \geq 0$ for any $A \in \mathcal{L}(\mathcal{H})$.

Let $\{P_i \in \mathcal{L}(\mathcal{H})\}_{i=1}^m$ be a set of projection operators. $\{P_i\}_{i=1}^m$ is called complete if $\sum_{i=1}^m P_i = \mathbb{I}$.

Exercise 70. Let $\{\psi_i \in \mathcal{H}\}_{i=1}^d$ be an orthonormal basis of $\mathcal{H}$. Prove that $\{|\psi_i\rangle\langle\psi_i|\}_{i=1}^d$ is a complete set of orthogonal projection operators, i.e., for $P_i := |\psi_i\rangle\langle\psi_i|$, it holds $P_i P_j = \delta_{ij} P_i$ and $\sum_i P_i = \mathbb{I}$. (Hint: use (A.1).)

Exercise 71. For a subspace of W of $\mathcal{H}$, define an operator P which maps $\psi \in \mathcal{H}$ to its closest vector ϕ in W as in Exercise 56. Show that P is a projection operator. (Hint: Use Exercise 59.) We say P is the projection operator **onto** the subspace W. Conversely, for any projection operator P, show that ran$P$ is the subspace in which P maps ψ to the closest vector.

Therefore, there exists one to one correspondence between projection operator P and its range $W = \text{ran}P$ in the above sense. In particular, projection operator P is called the **eigen projection** if the range of P is an eigenspace $E_A(\lambda)$.

Exercise 72. Let $\{\phi_i\}_{i=1}^m$ be an orthonormal basis of W, then show that the projection operator onto W can be written as $P = \sum_{i=1}^m |\phi_i\rangle\langle\phi_i|$.

[Spectral Decomposition and Eigenvalue Decomposition]

Exercise 73. Let A be a normal operator, and let λ be an eigenvalue of A with the corresponding eigenvector ψ. Prove that ψ is an eigenvector of $A^\dagger$ for an eigenvalue $\bar{\lambda}$. (Hint: show $||(A - \lambda)\psi||^2 = ||(A^\dagger - \bar{\lambda})\psi||^2$.)

Exercise 74. Prove that eigenvectors ψ and ϕ with distinct eigenvalues $\lambda_1 \neq \lambda_2$ of a normal operator A are orthogonal. [Hint: use Exercise 73 to show $(\lambda_1 - \lambda_2)\langle\psi, \phi\rangle = 0$.]

As a corollary, it turns out that eigenvectors corresponding to distinct eigenvalues are orthogonal for projection operators, positive operators, Hermitian operators, Unitary operators, etc.

Exercise 75. (Spectral Decomposition) Let A be an operator with the set of distinct eigenvalues $\{\lambda_i \in \mathbb{C}\}_{i=1}^m$ ($m \leq d$) and the corresponding eigenspaces $E_A(\lambda_i)$, with the spectral projection P_i. (i) Prove that the set $\{P_i\}$ is a complete set of orthogonal projection, if and only if A is a normal operator. (ii) As the corollary, show that any normal operator A can be written as

$$A = \sum_i \lambda_i P_i.$$

This form for normal operators (and hence for unitary operators, Hermitian operators, etc.) is called the **spectral decomposition** of A. (Hint: (ii) Simply check the following: $A\psi = A(\sum_i P_i)\psi = \sum_i AP_i\psi = \sum_i \lambda_i P_i\psi$ for any $\psi \in \mathcal{H}$.)

Since $\{P_i\}$ above is a complete set of orthogonal projection, one can construct an orthonormal basis $\{\psi_i\}_{i=1}^d$ of $\mathcal{H}$, by choosing orthonormal bases from each subspace $P_i\mathcal{H}$ (explain this in details). Then, we have

$$A = \sum_{j=1}^d \tilde{\lambda}_j |\psi_j\rangle\langle\psi_j| \tag{A.3}$$

where $\tilde{\lambda}_j$'s are all the eigenvalues of A where the degeneracy of A is taken into account. This decomposition is called an **eigenvalue decomposition**. If there is no confusion, $\tilde{\lambda}_i$ is denoted as λ_i.

Exercise 76. Show that an eigenvalue decomposition of A is not unique if there exist degenerate eigenvalues, while the spectral decomposition is always unique.

[Function of an operator] Let A be a normal operator with the spectral decomposition $A = \sum_i \lambda_i P_i$. Then, for any function $f : \mathbb{C} \to \mathbb{C}$ where the domain of f includes $\{\lambda_i\}$, the function $f(A)$ of A is defined as

$$f(A) := \sum_i f(\lambda_i) P_i. \tag{A.4}$$

For instance, for any positive operator $A \geq 0$, one can define $\sqrt{A} := \sum_i \sqrt{\lambda_i} P_i$. For the function $f(z) = |z|$ on $\mathbb{C}$, $|A|$ is defined by $|A| = \sum_i |\lambda_i| P_i$ for a normal operator. However, it is the convention for $|A|$ to extend this definition to any operator $A \in \mathcal{L}(\mathcal{H})$ by

$$|A| := \sqrt{A^\dagger A},$$

where the right hand side is the square root function $\sqrt{\ }$ of a positive operator $A^\dagger A$ (see Exercise 69). By definition, $|A|$ is positive for any $A \in \mathcal{L}(\mathcal{H})$. The eigenvalues of $|A|$ is called **singular values** of A.

Exercise 77. Let f be a polynomial function on $\mathbb{C}$: $f(z) := a_n z^n + a_{n-1} z^{n-1} + \cdots + a_1 z + a_0$ ($a_i \in \mathbb{C}$). Then, for a normal operator A, prove that

$$f(A) = a_n A^n + \cdots a_1 A + a_0 \mathbb{I} \tag{A.5}$$

according to the original definition (A.4).

Exercise 78. Prove that any spectral projection P_i of a normal operator A can be written in the form of polynomial (A.5). [Hint: Introduce the polynomials $f_i(z) := \frac{\Pi_{j \neq i}(z - \lambda_j)}{\Pi_{j \neq i}(\lambda_i - \lambda_j)}$ where $\{\lambda_i\}$ is the set of distinct eigenvalues of A. By noting that $f_i(\lambda_j) = \delta_{ij}$, and hence $P_i = f_i(A)$, using Exercise 77.]

Exercise 79. (Polar decomposition) Show that for any $A \in \mathcal{L}(\mathcal{H})$, there exists a unitary operator U such that $A = U|A|$. This is called the **polar decomposition** of A. [Hint: it is easy to see $\| |A|\psi \| = \|A\psi\|$, and thus one can construct a partial isometric operator W from ran $|A|$ to A, from which they have the same (finite) dimension. Then, one can easily construct U for the polar decomposition.]

[Trace Operation] A **trace** of an operator A is defined by

$$\operatorname{tr} A := \sum_{i=1}^{d} \langle \psi_i, A\psi_i \rangle,$$

where $\{\psi_i\}_{i=1}^{d}$ is an orthonormal basis of $\mathcal{H}$. Trace operation $\operatorname{tr} : \mathcal{L}(\mathcal{H}) \to \mathbb{C}$ is linear as is easy to see from the definition.

Exercise 80. Prove that the definition of the trace is independent of the choice of the orthonormal basis. [Hint: use Exercise 70.] In particular, show that $\operatorname{tr} A = \sum_{i=1}^{d} a_i$ where $\{a_i\}_{i=1}^{d}$ is the set of all the eigenvalues of A (Hint: when A is normal, use an eigenvalue decomposition (A.3). For a general operator, one might have to use the so-called Jordan decomposition.)

Exercise 81. (Cyclic property of trace) Show that the trace operation has a cyclic property, i.e., for any A and $B \in \mathcal{L}(\mathcal{H})$, it follows that

$$\operatorname{tr}[AB] = \operatorname{tr}[BA].$$

Exercise 82. For operator of the form $|\psi\rangle\langle\phi|$, show that $\operatorname{tr}[|\psi\rangle\langle\phi|A] = \langle\phi, A\psi\rangle$ for any $A \in \mathcal{L}(\mathcal{H})$. In particular, by putting $A = \mathbb{I}$, it follows $\operatorname{tr}[|\psi\rangle\langle\phi|] = \langle\phi, \psi\rangle$.

Exercise 83. Show that for any two positive operators A, B, $\operatorname{tr}[AB] \geq 0$. (Hint: use the spectral decomposition of A and linearity of tr.)

[Norms on $\mathcal{L}(\mathcal{H})$]

An **operator norm** of $A \in \mathcal{L}(\mathcal{H})$, denoted as $||A||_\infty$, is defined by

$$||A||_\infty := \sup_{\psi \in \mathcal{H},\, ||\psi||=1} ||A\psi||.$$

Exercise 84. Show that $||A||_\infty < \infty$ and this gives a norm on $\mathcal{L}(\mathcal{H})$.

A useful inner product between operators is the **Hilbert-Schmidt inner product** defined by

$$\langle A, B \rangle_{\mathrm{HS}} := \operatorname{tr}(A^\dagger B),$$

for $A, B \in \mathcal{L}(\mathcal{H})$.

Exercise 85. Prove that this is an inner product.

The **Hilbert-Schmidt norm**, denoted as $||A||_2$ is given by $||A||_2 := \sqrt{\langle A, A \rangle_{\mathrm{HS}}}$.

Another important norm, especially for density operators, is the **trace norm** denoted as $||A||_1$:

$$||A||_1 := \operatorname{tr}|A| = \operatorname{tr}\sqrt{A^\dagger A}.$$

Exercise 86. Prove that this gives a norm.

For positive operator A, it is obvious that $||A||_1 = \operatorname{tr} A$. In general, one can show

$$|\operatorname{tr} A| \leq ||A||_1, \tag{A.6}$$

for any $A \in \mathcal{L}(\mathcal{H})$.

Exercise 87. Using the Schwarz inequality for the Hilbert-Schmidt inner product and the polar decomposition, prove that $|\operatorname{tr}(AU)| \leq ||A||_1$ for any $A \in \mathcal{L}(\mathcal{H})$ and any unitary operator $U \in \mathcal{L}(\mathcal{H})$. Then, Eq. (A.6) follows by putting $U = \mathbb{I}$.

The indexes of the norms imply a unified treatment of these norms. Indeed, one can define a general norm, so-called Schatten p-norms, defined as

$$||A||_p := (\operatorname{tr}|A|^p)^{1/p} \tag{A.7}$$

for $1 \leq p \leq \infty$, where the case $p = \infty$ corresponds to the operator norm of A.

Using the singular values of A, it is easy to see $||A||_p = (\sum_{i=1}^m s_j^p)^{1/p}$ where $\{s_j\}_{j=1}^p$ is the set of all the singular values of A. Then, one can verify that $||A||_\infty = \max_j[s_j] = \lim_{p\to\infty} ||A||_p$ (show this), and this is why the operator norm is denoted with an index ∞. Using this, show the relation between the operator norm, the Hilbert-Schmidt norm, and the trace norm:

$$||A||_\infty \leq ||A||_2 \leq ||A||_1 \leq d||A||_\infty.$$

From this, the topologies on $\mathcal{L}(\mathcal{H})$ using any of the above norms the same. [Convex set] Let V be a (real) linear space. A subset W of V is called a **convex set** if W is closed under the **convex combination** operation, i.e., for all $\lambda \in [0,1], w_1, w_2 \in W, \lambda w_1 + (1-\lambda)w_2 \in W$.

Exercise 88. By illustrating concrete examples of convex sets in $\mathbb{R}^1, \mathbb{R}^2$ and $\mathbb{R}^3$, explain the geometrical meaning of a convex set.

Exercise 89. Let W be a convex set. Prove that for any $\{w_i \in W\}_{i=1}^n$ and a probability distribution $\{p_i \in \mathbb{R}\}_{i=1}^n$ ($p_i \geq 0, \sum_i p_i = 1$), a convex combination $\sum_i p_i w_i$ is an element of W.

[Extreme points] Let W be a convex set in V. An element w is called an **extreme point** of W if there is no nontrivial decompositions of w, i.e., if $w = \lambda w_1 + (1-\lambda)w_2$ for some $\lambda \in (0,1)$ and $w_1, w_2 \in W$, then $w = w_1 = w_2$.

Exercise 90. By illustrating concrete examples of convex sets in $\mathbb{R}^1, \mathbb{R}^2$ and $\mathbb{R}^3$, explain the geometrical meaning of an extreme point.

Appendix B. Solutions

Solutions of all the exercises and errata will be uploaded in the web page: http://staff.aist.go.jp/gen-kimura/decoherence.pdf

References

1. M. A. Nielsen and I. L. Chuang: *Quantum Computation and Quantum Information* (Cambridge University Press, Cambridge, 2000).
2. D. Giulini, E. Joos, C. Kiefer, J. Kupsch, I. O. Stamatescu and H. D. Zeh: *Decoherence and the Appearance of a Classical World in Quantum Theory* (Berlin: Springer; 1996, second revised edition, 2003); The draft for second edition of Chapter 2 by H. D. Zeh is available in e-print quant-ph/9506020.
3. H. D. Zeh: Found. Phys. **1** (1970) 69; **3** (1973) 109.
4. W. H. Zurek: Physical Review D **24** (1981) 1516; **26** (1982) 1516.
5. W. H. Zurek: Physics Today **44** (1991) 36; updated version is available in e-print quant-ph/0306072.
6. G. Bacciagaluppi: "The Role of Decoherence in Quantum Mechanics" in the Stanford Encyclopedia of Philosophy <http://plato.stanford.edu/entries/qm-decoherence/>.
7. N. G. van Kampen: *Stochastic Processes In Physics And Chemistry* (North-Holland, 1983).
8. D. F. Walls and G. J. Milburn: *Quantum Optics* (Springer-Verlag, Heidelberg, 1994); M. O. Scully and M. S. Zubairy: *Quantum Optics* (Cambridge University Press, Cambridge, 1997).
9. C. W. Gardiner and P. Zoller: *Quantum Noise*, 2nd ed. (Springer-Verlag, Heidelberg, 2000).
10. H. P. Breuer and F. Petruccione: *The theory of open quantum systems* (Oxford, 2002).
11. E. B. Davies: *Quantum Theory of Open Systems* (Academic Press, London, 1976).
12. R. Alicki and K. Lendi: *Quantum Dynamical Semigroups and Applications*, Lecture Notes in Physics Vol. 286 (Springer, Berlin, 1987).
13. K. Kraus: Ann. Phys. **64** (1971) 311; K. Kraus: *States, Effects, and Operations* (Springer, Berlin, 1983).
14. R. F. Streater: *Statistical Dynamics — A stochastic Approach to Nonequilibrium Thermodynamics* (Imperial College, London, 1995).
15. P. A. M. Dirac: *The Principles of Quantum Mechanics* (Oxford, 1958).

16. A. Peres: *Quantum Theory: Concepts and Methods* (Kluwer Academic, London, 1998)
17. B. d'Espagnat: *Conceptual Foundations of Quantum Mechanics*, 2nd ed. (Addison-Wesley, 1976)
18. GG. Ludwig: *Foundations of Quantum Mechanics I* (Springer, Berlin, 1983).
19. H. Nagaoka: Trans. Jpn. Soc. Ind. Apple. Math., 1, 43 (1991).
20. M. Hayashi: *Quantum Information: An Introduction* (Springer, Berlin, 2006).
21. C. Tsallis: J. Stat. Phys. **52** (1988) 479; S. Abe and Y. Okamoto: *Nonextensive Statistical Mechanics and Its Applications* (Springer, Berlin, 2000).
22. F. Bloch: Phys. Rev. **70** (1946) 460.
23. F. T. Hioe and J. H. Eberly: Phys. Rev. Lett. **47** (1981) 838.
24. G. Mahler and V. A. Weberruss: *Quantum Networks* (Springer, Berlin, 1995).
25. L. Jakóbczyk and M. Siennicki: Phys. Lett. A **286** (2001) 383.
26. G. Kimura: Phys. Lett. A **314** (2003) 339; G. Kimura and A. Kossakowski: Open Sys. Information Dyn. **12** (2005) 207.
27. W. Heisenberg: Z. Phys. **43** (1927) 172.
28. M. Ozawa: Ann. Phys. **311** (2004) 350.
29. T. Miyadera and H. Imai: Phys. Rev. A **78** (2008) 052119.
30. M. Ozawa: Rep. on Math. Phys. **18** (1980) 11. See also Theorem II.2 in Ref. 28.
31. A. S . Holevo: *Probabilistic and Statistical Aspects of Quantum Theory* (North-Holland, Amsterdam, 1982).
32. E. B. Davies and J. T. Lewis: Comm. Math. Phys. **17** (1970) 239.
33. M. Ozawa: J. Math. Phys. **25** (1984) 79.
34. J. von Neumann: *Mathematische Grundlagen der Quantenmechanik* (Springer, Berlin, 1932).
35. R. Haag: *Local Quantum Physics: Fields, Particles, Algebras* (Springer, Berlin, 1991); H. Araki: *Mathematical Theory of Quantum Fields* (Oxford University Press, 2000).
36. R. Bhatia: *Matrix Analysis* (Springer, Berlin, 1997).
37. E. B. Davies and Z. Wahrscheinlichkeitstheorie: **23** (1972) 161; D. E. Evans: Commun. Math. Phys. **48** (1976) 15.
38. K. Yoshida: *Functional Analysis* (Springer, Berlin, 1972).
39. A. Kossakowski: Rep. Math. Phys **3** (1972) 247; A. Kossakowski: Bull. Acad. Pol. Sci. Ser. Math. Astr. Phys. **21** (1972) 1021; R. S. Ingarden and A. Kossakowski: Ann. Phys. **89** (1975) 451.
40. G. Lindblad: Commun. Math. Phys. **48** (1976) 119.
41. V. Gorini, A. Kossakowski and E. C. G. Sudarshan: J. Math. Phys. **17** (1976) 821.

QUANTUM ERROR CORRECTION
AND
FAULT-TOLERANT QUANTUM COMPUTING

FRANK GAITAN

Department of Physics, Southern Illinois University
Carbondale, IL 62901-4401, USA
Advanced Sciences Institute, The Institute of Physical and Chemical Research
(RIKEN)
Wako-shi, Saitama 351-0198, JAPAN
CREST, Japan Science and Technology Agency (JST)
Kawaguchi, Saitama 332-0012, JAPAN
E-mail: gaitan@physics.siu.edu

RAN LI

Department of Physics, Kent State University, Stark Campus
North Canton, OH 44720, USA
Advanced Sciences Institute, The Institute of Physical and Chemical Research
(RIKEN)
Wako-shi, Saitama 351-0198, JAPAN
CREST, Japan Science and Technology Agency (JST)
Kawaguchi, Saitama 332-0012, JAPAN

We review the theories of quantum error correction, and of fault-tolerant quantum computing, and show how these powerful tools are combined to prove the accuracy threshold theorem for a particular error model. One of the theorem's assumptions is the availability of a universal set of unencoded quantum gates whose error probabilities P_e fall below a value known as the accuracy threshold P_a. For many, $P_a \sim 10^{-4}$ has become a rough estimate for the threshold so that quantum gates are anticipated to be approaching the accuracies needed for fault-tolerant quantum computing when $P_e < 10^{-4}$. We show how controllable quantum interference effects that arise during a type of non-adiabatic rapid passage can be used to produce a universal set of quantum gates whose error probabilities satisfy $P_e < 10^{-4}$. We close with a discussion of the current challenges facing an experimental implementation of this approach to reliable universal quantum computation.

Keywords: Quantum Error Correction; Fault-Tolerant Quantum Computing; Accuracy Threshold Theorem; High-Fidelity Universal Quantum Gates

1. Introduction

Over a surprisingly short period of time (1994-1997), quantum computing evolved from a field many regarded as largely of academic interest, to
one undergoing enormous growth driven by the sober expectation that a
fully-functioning, scalable quantum computer can be built. What led to
this remarkable reversal? First, quantum algorithms for factoring, discrete-
log, and searching appeared in the literature which demonstrated the computational power inherent in a quantum computer. Second, the theory of
quantum error correction was established which showed that errors arising during a quantum computation could be detected and removed. Third,
fault-tolerant protocols were introduced that controlled the way errors could
spread during a quantum computation. The culmination of these three lines
of development was the proof of the accuracy threshold theorem which
showed that, under appropriate conditions, an arbitrarily long quantum
computation could be done with arbitrarily small error probability in the
presence of noise and using imperfect quantum gates. The theorem established that no physical law stood in the way of building a quantum computer. The various proofs of the theorem made it clear that the task would
be technically challenging, but not impossible. This came as a surprise to
many, and set in motion the enormous research effort alluded to above.

In this chapter we review the theories of quantum error correction
(Sec. 2), and of fault-tolerant quantum computing (Sec. 3). We then show
how these powerful tools are combined to prove the accuracy threshold
theorem for a particular error model (Sec. 4). Due to space limitations, we
cannot present the arguments in Secs 2–4 in full detail. Readers interested
in seeing the complete arguments are referred to Ref. 1 where they are
given a book-length exposition. Finally, we show how controllable quantum
interference effects (arising during a special type of non-adiabatic rapid
passage) can be used to implement a high-fidelity universal set of quantum gates whose performance approaches the requirements of the accuracy
threshold theorem (Sec. 5). We close with a discussion of the current challenges facing an experimental implementation of this approach to reliable
universal quantum computation.

2. Quantum Error Correction

In this section we present the theory of quantum error correcting codes
(QECCs). We begin by introducing a number of ideas from the theory of
classical error correcting codes that, upon generalization, are incorporated

into the theory of QECCs (Sec. 2.1). We define QECCs in Sec. 2.2, and introduce quantum stabilizer codes in Sec. 2.3. Finally, we describe how concatenated QECCs are constructed in Sec. 2.4. Concatenated codes based on quantum stabilizer codes will turn out to be an essential ingredient in the proof of the accuracy threshold theorem given in Sec. 4.

2.1. *Classical Error Correcting Codes*

The theory of classical error correcting codes is a beautiful, highly developed subject.[2] We will focus on linear block codes as they are the most relevant for QECCs. To introduce these codes, consider the problem of noisy communication. Here the aim is to send a stream of encoded symbols through a noisy channel as reliably as possible. The message symbols are first encoded into a block of k bits $\mathbf{m} = (m_1, \ldots, m_k)$ with $m_i = 0, 1$. The 2^k message symbols $\mathbf{m}$ are the elements of the k-dimensional vector space $F_2^k = F_2 \otimes \cdots \otimes F_2$ over the finite field $F_2 = 0, 1$. Addition and multiplication in F_2 are modulo 2. For transmission through the channel, the message symbols $\mathbf{m}$ are further encoded into 2^k codewords $\mathbf{c} = (c_1, \ldots, c_n)$ which are elements of the n-dimensional vector space F_2^n. (We will always represent a vector $\mathbf{v}$ by a column vector and its transpose $\mathbf{v}^T$ by a row vector.) The encoding $\mathbf{m} \to \mathbf{c}$ is defined to be a linear map implemented by an $n \times k$ matrix G known as the generator matrix: $\mathbf{c} = G\,\mathbf{m}$. It follows that the codewords $\{\mathbf{c}\}$ belong to the k-dimensional subspace $\mathcal{C}$ in F_2^n spanned by the columns of G. A linear block code $\mathcal{C}$ that maps k bits into n bits is said to be an (n,k) code.

It proves useful to introduce an inner product on F_2^n. If $\mathbf{u}, \mathbf{v} \in F_2^n$, the inner product $\mathbf{u} \cdot \mathbf{v}$ is defined to be

$$\mathbf{u} \cdot \mathbf{v} = \sum_{i=1}^{n} u_i v_i \quad (\mathrm{mod}\ 2)$$

which takes values in $F_2 = 0, 1$. Two vectors $\mathbf{u}$ and $\mathbf{v}$ are said to be orthogonal if $\mathbf{u} \cdot \mathbf{v} = 0$. Note that a vector $\mathbf{v} \in F_2^n$ can be self-orthogonal: $\mathbf{v} \cdot \mathbf{v} = 0$.

A block code can also be defined by imposing linear constraints known as parity checks on all codewords. Each linear constraint is specified by introducing a parity check vector $\mathbf{h} \in F_2^n$, and demanding that all codewords be orthogonal to it: $\mathbf{h} \cdot \mathbf{c} = 0$. To construct an (n,k) block code $\mathcal{C}$, $n - k$ linearly independent parity checks are introduced $\{\mathbf{h}_i \cdot \mathbf{c} = 0 \mid i = 1, \ldots, n - k\}$ so that each codeword is left with k degrees of freedom. The codewords thus form a k-dimensional subspace $\mathcal{C}$ in F_2^n which is identified with the code.

The row vectors $\mathbf{h}_i^T$ are gathered together to form the rows of an $(n-k) \times n$ matrix H known as the parity check matrix:

$$H = \begin{pmatrix} -\!\!-\!\!- & \mathbf{h}_1^T & -\!\!-\!\!- \\ & \vdots & \\ -\!\!-\!\!- & \mathbf{h}_{n-k}^T & -\!\!-\!\!- \end{pmatrix}. \tag{1}$$

By construction, the parity check matrix H annihilates all codewords $H\mathbf{c} = 0$. The $(n-k)$–dimensional subspace $\mathcal{C}^\perp$ spanned by the $\{\mathbf{h}_i\}$ constitutes an $(n, n-k)$ block code known as the dual code $\mathcal{C}^\perp$ of $\mathcal{C}$. From eq. (1), $\mathcal{C}^\perp$ corresponds to the column-space of H^T. It follows from the definition of the parity check vectors $\{\mathbf{h}_i\}$ that the codewords of $\mathcal{C}^\perp$ are orthogonal to the codewords of $\mathcal{C}$, and that H^T is the generator matrix for $\mathcal{C}^\perp$. It is straightforward to show that the parity check matrix for $\mathcal{C}^\perp$ is G^T.[1,2] Note that since a vector $\mathbf{v} \in F_2^n$ can be self-orthogonal, it is possible for a codeword in $\mathcal{C}$ to also belong to the dual code $\mathcal{C}^\perp$! A code $\mathcal{C}$ is said to be self-dual if $\mathcal{C} = \mathcal{C}^\perp$, and weakly self-dual if $\mathcal{C} \subset \mathcal{C}^\perp$.

Returning to the problem of noisy communication, suppose the codeword $\mathbf{c}$ is the input to the communication channel and $\mathbf{y} \in F_2^n$ is the channel output. Because of noise in the channel, the output $\mathbf{y}$ may differ from the input $\mathbf{c}$. We define the (transmission) error $\mathbf{e}$ to be the difference between the input and output vectors $\mathbf{e} = \mathbf{y} - \mathbf{c}$. The challenge for the decoder at the receiving end of the channel is to recover the input codeword $\mathbf{c}$ from the channel output $\mathbf{y}$ with high probability. A number of key ideas must be introduced before we can discuss sensible decoding strategies.

We begin with the Hamming weight $wt(\mathbf{v})$ of a vector $\mathbf{v} \in F_2^n$ which is defined to be the number of its non-zero components

$$wt(\mathbf{v}) = \sum_{i=1}^{n} v_i.$$

The Hamming weight allows a distance function to be defined on F_2^n. Let $\mathbf{u}$ and $\mathbf{v}$ be vectors in F_2^n. We define the Hamming distance $d(\mathbf{u}, \mathbf{v})$ between $\mathbf{u}$ and $\mathbf{v}$ to be the Hamming weight of $\mathbf{u} - \mathbf{v}$:

$$d(\mathbf{u}, \mathbf{v}) = wt(\mathbf{u} - \mathbf{v}).$$

Now suppose that we have a (block) code $\mathcal{C}$, and $\mathbf{c}$ and $\mathbf{c}'$ are two codewords. Since $\mathcal{C}$ is a subspace of F_2^n, $\mathbf{c}'' = \mathbf{c} - \mathbf{c}'$ also belongs to the subspace $\mathcal{C}$ and so is a codeword. Then

$$d(\mathbf{c}, \mathbf{c}') = wt(\mathbf{c} - \mathbf{c}') = wt(\mathbf{c}''). \tag{2}$$

The minimum code distance d is defined to be

$$d = \min_{\mathbf{c} \neq \mathbf{c}'} d(\mathbf{c}, \mathbf{c}') = \min_{\mathbf{c}'' \neq 0} wt(\mathbf{c}''),$$

where $\mathbf{c}$, $\mathbf{c}'$, and $\mathbf{c}''$ are codewords. Thus the minimum code distance d is the smallest non-zero weight of the codewords. If the $\mathcal{C}$ maps k bits into n bits and has minimum code distance d, we say it is an (n, k, d) code, slightly extending our earlier notation.

A simple model for a noisy communication channel is the binary symmetric channel (BSC). For this channel: the probability that a received bit is in error (correct) is p $(1 - p)$; errors on different received bits are uncorrelated; and the channel output is a stationary random process. For such a channel, the probability that t erroneous bits are received is p^t.

A useful decoding scheme for the BSC is nearest neighbor decoding. In this scheme, if $\mathbf{y}$ is the channel output, the decoder returns the codeword $\mathbf{c}_*$ that is the smallest Hamming distance from $\mathbf{y}$. Thus $d(\mathbf{c}_*, \mathbf{y}) \leq d(\mathbf{c}, \mathbf{y})$ for all $\mathbf{c} \in \mathcal{C}$. The rationale for this decoding scheme is that the best-guess error $\mathbf{e}_* = \mathbf{y} - \mathbf{c}_*$ has the smallest possible weight. Consequently, $\mathbf{e}_*$ has the smallest number of bit errors, and so is a most probable error. If more than one error has the same weight as e_*, then a decoding error is still possible. It can be shown that nearest neighbor decoding minimizes the probability of a decoding error when codewords are equally likely to be sent.[2]

Let H be the parity check matrix for the code $\mathcal{C}$ and $\mathbf{y}$ a possible channel output. The syndrome $S(\mathbf{y})$ of $\mathbf{y}$ is defined to be $S(\mathbf{y}) = H \mathbf{y}$. If $\mathbf{c}$ was the channel input so that $\mathbf{y} = \mathbf{c} + \mathbf{e}$, then

$$S(\mathbf{y}) = H \mathbf{y} = H(\mathbf{c} + \mathbf{e}) = H \mathbf{e} = S(\mathbf{e}),$$

where we used that H annihilates codewords $\mathbf{c}$. Thus the error $\mathbf{e}$ and the channel output $\mathbf{y}$ have the same error syndrome. Suppose that $\mathcal{C}$ is an (n, k, d) code. We define the coset $\mathbf{y} + \mathcal{C}$ of $\mathcal{C}$ containing $\mathbf{y} \in F_2^n$ to be the set

$$\mathbf{y} + \mathcal{C} = \{ \mathbf{y} + \mathbf{c} \,|\, \mathbf{c} \in \mathcal{C} \}.$$

Since a vector space is an Abelian group under vector addition, the cosets of $\mathcal{C}$ partition F_2^n:

$$F_2^n = \mathcal{C} \cup (l_1 + \mathcal{C}) \cup \cdots \cup (l_m + \mathcal{C}), \tag{3}$$

where $m = 2^{n-k} - 1$. In the coset $l_i + \mathcal{C}$, the vector l_i is called the coset leader, and it is the smallest weight element of this coset. If there is more than one smallest weight element, one arbitrarily chooses one of these elements to be the coset leader.

In standard array decoding, if $\mathbf{y}$ is the channel output, the decoding protocol is to find the coset that contains $\mathbf{y}$, identify its coset leader $\mathbf{l}$, and return $\mathbf{c}_* = \mathbf{y} - \mathbf{l} \in C$ as the best-guess input codeword. For a BSC, $\mathbf{l}$ will be a most probable error, and $\mathbf{c}_*$ a most probable input codeword. If $\mathbf{c}_*$ is not the actual input codeword, then a decoding error occurs. In standard array decoding, a decoding error occurs if and only if the error $\mathbf{e}$ is not a coset leader.[1] Let α_i be the number of coset leaders of weight i, and let p be the bit error probability for a BSC. Then the probability that $\mathbf{e}$ is a coset leader is $\alpha_i p^i (1-p)^{n-i}$ and the probability that decoding succeeds is

$$P_{suc} = \sum_{i=0}^{n} \alpha_i p^i (1-p)^{n-i}.$$

Since decoding either succeeds or fails, it follows that the probability for a decoding error occurs is

$$P_e = 1 - \sum_{i=0}^{n} \alpha_i p^i (1-p)^{n-i}. \tag{4}$$

If $P_e < p$, then using C to encode message symbols for transmission over a noisy communication channel improves the reliability of the channel.

2.2. Quantum Error Correcting Codes

The quantum analog of a classical bit is a qubit whose state space is the two-dimensional Hilbert space H_2. We will usually use the eigenstates of σ_z as the single-qubit computational basis (CB) states $|0\rangle$ and $|1\rangle$: $\sigma_z |i\rangle = (-1)^i |i\rangle$, where $i = 0, 1$. For a quantum register containing n qubits, the state space is the n-qubit Hilbert space $H_2^n = H_2 \otimes \cdots \otimes H_2$ which is the direct product of n single-qubit Hilbert spaces H_2. The n-qubit CB states are constructed by forming all possible direct products of the single-qubit CB states: $|i_1 \cdots i_n\rangle = |i_1\rangle \otimes \cdots \otimes |i_n\rangle$, where $i_1, \ldots, i_n = 0, 1$. It follows that H_2^n is 2^n-dimensional.

A quantum error correcting code (QECC) that encodes k qubits into n qubits is defined through an encoding map ξ from the k-qubit Hilbert space H_2^k onto a 2^k-dimensional subspace C_q of the n-qubit Hilbert space H_2^n. Just as with a classical linear code, we identify a QECC with the image space C_q. The encoding operation $\xi : H_2^k \to C_q$ is required to be unitary and so it establishes a 1-1 correspondence between the unencoded and encoded CB states $|i\rangle \equiv |i_1 \cdots i_k\rangle$ and $|\bar{i}\rangle \equiv |\overline{i_1 \cdots i_k}\rangle$, respectively:

$$|\overline{i_1 \cdots i_k}\rangle = \xi |i_1 \cdots i_k\rangle. \tag{5}$$

Thus, $|\bar{0}\rangle$ is the encoded image of $|0\rangle$, and $|\overline{011}\rangle$ is the encoded image of $|011\rangle$. ξ also maps unencoded single-qubit operators o^j into encoded qubit operators O^j: $o^j \rightarrow O^j = \xi o^j \xi^\dagger$, where $j = 1, \ldots, k$ labels the qubits. For example, ξ maps $\sigma_z^j \rightarrow Z_j = \xi \sigma_z^j \xi^\dagger$. Not surprisingly, the encoded CB states $\{|\bar{i}\rangle\}$ are simultaneous eigenstates of the $\{Z_j\}$: $Z_j|\bar{i}\rangle = (-1)^{i_j}|\bar{i}\rangle$, and the encoding preserves the eigenvalue $(-1)^{i_j}$. The states $|c\rangle \in \mathcal{C}_q$ are the codewords, and since the encoded CB states $\{|\bar{i}\rangle\}$ span $\mathcal{C}_q$, we can write $|c\rangle = \sum_i c_i |\bar{i}\rangle$.

The dynamical effects produced by the environment of a quantum register on the register's state can be modeled by a set of linear operators $E = \{E_a\}^4$ which are the errors that the environment can introduce into a quantum computation. The following theorem establishes necessary and sufficient conditions that must be satisfied by a QECC if it is to correct a given set of errors.

Theorem 2.1. *Let $\mathcal{C}_q$ be a QECC and $E = \{E_a\}$ a set of environmental errors. $\mathcal{C}_q$ will correct the errors in E if and only if for all encoded CB states $|\bar{i}\rangle$, $|\bar{j}\rangle$ and error operators $E_a, E_b \in E$:*

$$\langle \bar{i}|E_a^\dagger E_b|\bar{j}\rangle = C_{ab}\,\delta_{ij}. \tag{6}$$

Note that the matrix C formed from the constants C_{ab} is Hermitian. For a proof of this theorem, see Refs. 3 and 1. If the eigenvalues c_i of C are strictly positive, then $\mathcal{C}_q$ is said to be a non-degenerate QECC; otherwise it is a degenerate code.

It is important to note that a linear combination of correctable errors is also a correctable error. This was established in Ref. 3 where it was shown that the maximal set of errors correctable by a QECC is linearly closed. Thus it is possible to focus on correcting a basis set of errors. To generate such a set, recall that the errors I, σ_x, σ_y, and σ_z form a basis for single-qubit errors, and the basis for n-qubit errors is produced from them by forming all possible n-fold direct products:

$$e(j_1, \ldots, j_n) = \sigma_{j_1}^1 \otimes \cdots \otimes \sigma_{j_n}^n. \tag{7}$$

Here superscripts on the RHS label the qubits $1, \ldots, n$; $j_i = 0, x, y, z$; and σ_0^i is the identity operator I^i for the i-th qubit. This n-qubit error basis can be transformed into a multiplicative group known as the Pauli group $\mathcal{G}_n$ if we allow the elements $e(j_1, \ldots, j_n)$ to be multiplied by -1 and $\pm i$. This is necessary since products such as $\sigma_x \sigma_y = i\sigma_z$ generate factors of i and closure under multiplication requires $i\sigma_z \in \mathcal{G}_n$. The weight of an error operator $e(j_1, \ldots, j_n) \in \mathcal{G}_n$ is defined to be the number of qubits on which $\sigma^i_{j_i} \neq I^i$. Thus, the error operator $\sigma^1_x \otimes I^2 \otimes \sigma^3_y$ has weight 2.

The condition for quantum error correction, eq. (6), allows us to define the distance of a QECC. Writing $E = E^\dagger_a E_b$, eq. (6) becomes

$$\langle \bar{i} | E | \bar{j} \rangle = C_E \delta_{ij}, \tag{8}$$

where all the dependence on the CB states on the RHS resides in the Kronecker delta δ_{ij}. A QECC has distance d if all errors of weight less than d satisfy eq. (8), and at least one error of weight d exists that violates it. The notation $[n, k, d]$ is usually used to denote a QECC that maps k qubits into n qubits and which has distance d. It is possible to show (see Ref. 1) that if $\mathcal{C}_q$ corrects (detects) t errors, then $d = 2t + 1$ ($t + 1$).

Entanglement[5] can be used to detect and correct errors that may arise during a quantum computation. We describe the basic protocol here. A discussion of how this protocol is implemented fault-tolerantly will be given in Sec. 3.2. Assume that initially our quantum computer Q and its environment E are not entangled, and that Q's initial state is $|\chi\rangle$ while that of E is $|e\rangle$. Due to the unavoidable coupling between Q and E, the two systems become entangled:

$$|e\rangle |\chi\rangle \longrightarrow \sum_s |e_s\rangle \{ E_s |\chi\rangle \}. \tag{9}$$

The final states of the environment $\{|e_s\rangle\}$ need not form an orthonormal set. The operators $\{E_s\}$ are the environmental errors described above. At this point in the protocol, ancilla qubits are introduced that are coupled to the qubits in Q. The ancilla are initially prepared in the fiducial state $|a_0\rangle$. The unitary interaction U that couples them to Q is designed to produce the following effect:

$$U [|a_0\rangle \{ E_s |\chi\rangle \}] = |s\rangle \{ E_s |\chi\rangle \}. \tag{10}$$

The final ancilla state $|s\rangle$ is assumed to depend on E_s, but not on $|\chi\rangle$. The ancilla states $\{|s\rangle\}$ are assumed to form an orthonormal set. Because the quantum dynamics is linear, U produces the following transformation:

$$U \left[\sum_s |e_s\rangle |a_0\rangle \{ E_s |\chi\rangle \} \right] = \sum_s |e_s\rangle |s\rangle \{ E_s |\chi\rangle \}. \tag{11}$$

Since the $\{|s\rangle\}$ are orthonormal, they can be distinguished.[1] If we measure the ancilla in the s-basis, and find the result S, then the post-measurement state is the *non-entangled* state $|\psi_{pm}\rangle$:

$$|\psi_{pm}\rangle = |e_S\rangle |S\rangle \{ E_S |\chi\rangle \}. \tag{12}$$

S plays the role of error syndrome in quantum error correction, and the process of determining it through measurement of the ancilla is known as syndrome extraction. Since S is known from the ancilla measurement, and it is assumed that E_S can be identified from S (more on this below), we can apply E_S^{-1} to Q, leaving the composite system in the final state:

$$|e_S\rangle|S\rangle|\chi\rangle. \tag{13}$$

This sequence of operations and measurements has: (1) unentangled the quantum computer Q from its environment E and from the ancilla; and (2) put Q back into its initial state $|\chi\rangle$ (viz. removed the errors introduced by the environment). Thus the unwanted entanglement between the quantum computer and its environment has been removed by further entangling the quantum computer with a set of ancilla qubits. Appropriate measurement of the ancilla then removes the unwanted entanglement and indicates what operation must be applied to Q to correct errors (see below).

Two comments are in order before moving on to our discussion of quantum stabilizer codes. (1) It is not necessary for the error syndrome S to identify E_S. It is enough if S allows us to identify an operation that corrects the quantum computer's state and disentangles it from its environment. (2) In general, it will not be possible to correct all the errors E_s. Error correcting codes, classical or quantum, are only able to fix a definite set of errors. For useful codes, the correctable errors are also the most probable errors.

2.3. Quantum Stabilizer Codes

Having discussed $[n, k, d]$ QECCs in general, we now focus on the important class of codes known as quantum stabilizer codes[6-8]. For these codes the code space $\mathcal{C}_q$ is identified with the unique subspace of H_2^n that is fixed by the elements of an Abelian group $\mathcal{S}$ known as the stabilizer of $\mathcal{C}_q$. Specifically, for all stabilizer elements $s \in \mathcal{S}$ and all codewords $|c\rangle \in \mathcal{C}_q$:

$$s|c\rangle = |c\rangle. \tag{14}$$

The stabilizer $\mathcal{S}$ is constructed from a set of $n - k$ operators $g_1, \ldots, g_{n-k}$ known as the generators of $\mathcal{S}$. Each element $s \in \mathcal{S}$ can be written as a unique product of powers of the generators:

$$s = g_1^{p_1} \cdots g_{n-k}^{p_{n-k}}. \tag{15}$$

Because $\mathcal{S}$ is Abelian, the generators are a mutually commuting set of operators. By definition, the stabilizer $\mathcal{S}$ is a subgroup of the Pauli group

$\mathcal{G}_n$ and so each generator g_j can be written as (see eq. (7)):

$$g_j = \sigma_{j_1}^1 \otimes \cdots \otimes \sigma_{j_n}^n, \tag{16}$$

where $j_i = 0, x, y, z$. Since $\sigma_y^k = i\sigma_z^k\sigma_x^k$, we can factor each σ_y in eq. (16) and then separate the σ_x factors from the σ_z factors so that eq. (16) becomes

$$g_j = i^{\lambda_j} \sigma_x(a_j)\sigma_z(b_j)$$
$$\equiv i^{\lambda_j} \left[\prod_{k=1}^n \left(\sigma_x^k\right)^{a_{j,k}} \right] \left[\prod_{l=1}^n \left(\sigma_z^l\right)^{b_{j,l}} \right], \tag{17}$$

where $a_j = (a_{j,1}, \ldots, a_{j,n})$ and $b_j = (b_{j,1}, \ldots, b_{j,n})$. It follows from eq. (17) that $g_j^2 = I^j$ and so each generator has order 2. The exponent λ_j is chosen so that g_j is Hermitian for all $j = 1, \ldots, n-k$. From these remarks, we see that the generators are also unitary and have eigenvalues $\Lambda_j = \pm 1$. Finally, because the generators have order 2, eq. (15) establishes that: (i) for each j, $p_j = 0, 1$; and (ii) $\mathcal{S}$ has 2^{n-k} elements.

It is a simple matter to show[1] that if $e, f \in \mathcal{G}_n$, then either e and f commute ($[e, f] = 0$) or anticommute ($\{e, f\} = 0$). Since the generators $g_j \in \mathcal{G}_n$, it follows that for any $e \in \mathcal{G}_n$ and for all $j = 1, \ldots, n-k$, either $[e, g_j] = 0$ or $\{e, g_j\} = 0$. This allows us to define the syndrome $S(e)$ of the error $e \in \mathcal{G}_n$ as follows.

Definition 2.1. *Let $\mathcal{C}_q$ be a quantum stabilizer code with generators $g_1, \ldots, g_{n-k}$ and let e be an arbitrary error in the Pauli group $\mathcal{G}_n$. The syndrome $S(e)$ of e is the bit string $l = l_1 \cdots l_{n-k}$, where the component bits l_i are determined by*

$$l_i = \begin{cases} 0 & \text{if } [e, g_i] = 0 \\ 1 & \text{if } \{e, g_i\} = 0 \end{cases} \quad (i = 1, \ldots, n-k). \tag{18}$$

The error syndrome allows a number of useful results for $\mathcal{C}_q$ to be established.[1]

(1) It follows from Definition 2.1 that if e has a non-vanishing error syndrome $S(e) \neq 0 \cdots 0 \equiv 0$, it must anticommute with a subset of the generators of $\mathcal{S}$. Let g_k be one of these generators and $|i\rangle, |j\rangle$ arbitrary encoded CB states (basis codewords); then

$$\langle i|e|j\rangle = \langle i|eg_k|j\rangle = -\langle i|g_ke|j\rangle = -\langle i|e|j\rangle,$$

where we used that g_k fixes all codewords. It follows that $\langle i|e|j\rangle = 0$ for an error with non-vanishing error syndrome. Thus $e|j\rangle$ is orthogonal to $|i\rangle$ and so the erroneous basis codeword $e|j\rangle$ is distinguishable from the basis codeword $|i\rangle$, for all basis codewords $|i\rangle$ and $|j\rangle$. Thus $\mathcal{C}_q$ can detect all errors $e \in \mathcal{G}_n$ with non-vanishing error syndrome.

(2) Let $E = \{E_a\}$ be a set of errors belonging to $\mathcal{G}_n$ for which $S(E_a^\dagger E_b) \neq 0$ for all $E_a, E_b \in E$. Repeating the argument given in (1), we have that $\langle i|E_a^\dagger E_b|j\rangle = 0$, for all basis codewords $|i\rangle$ and $|j\rangle$. This is just eq. (6) with $C_{ab} = 0$ and so the set of errors E is correctable by $\mathcal{C}_q$.

(3) Errors having a vanishing error syndrome $S(e) = 0$ commute with all generators of $\mathcal{S}$. The centralizer $\mathcal{C}(\mathcal{S})$ of $\mathcal{S}$ is the set of errors $e \in \mathcal{G}_n$ that commute with all elements of $\mathcal{S}$. Since the stabilizer $\mathcal{S}$ is Abelian, $\mathcal{S} \subset \mathcal{C}(\mathcal{S})$. Let $e \in \mathcal{C}(\mathcal{S})$. Either (i) $e \in \mathcal{S}$, or (ii) $e \in \mathcal{C}(\mathcal{S}) - \mathcal{S}$, the complement of $\mathcal{S}$ in $\mathcal{C}(\mathcal{S})$. In the first case e is detectable and trivially correctable since it fixes codewords and so is really no error at all. For the second case, $e \in \mathcal{C}(\mathcal{S}) - \mathcal{S}$, it is possible to show[1] that this set of errors is non-detectable by $\mathcal{C}_q$. The essential point is that for each basis codeword $|i\rangle$ and $e \in \mathcal{C}(\mathcal{S}) - \mathcal{S}$, there exists at least one different basis codeword $|k_*\rangle \neq |i\rangle$ for which $\langle k_*|e|i\rangle \neq 0$. Thus $e|i\rangle$ is not orthogonal to $|k_*\rangle$ and so the two states cannot be distinguished by $\mathcal{C}_q$. Thus $\mathcal{C}_q$ cannot distinguish between an erroneous basis codeword $(e|i\rangle)$ and an error-free basis codeword $(|k_*\rangle)$ when $e \in \mathcal{C}(\mathcal{S}) - \mathcal{S}$, and so $\mathcal{C}_q$ is unable to detect such errors.

(4) Ref. 1 establishes the following two properties. (a) A quantum stabilizer code $\mathcal{C}_q$ has distance d if and only if $\mathcal{C}(\mathcal{S}) - \mathcal{S}$ has an element of weight d and none of weight less than d (other than the identity element). (b) A quantum stabilizer code $\mathcal{C}_q$ is degenerate if and only if its stabilizer $\mathcal{S}$ has an element of weight less than d (besides the identity element).

(5) Since the centralizer $\mathcal{C}(\mathcal{S})$ is a subgroup of $\mathcal{G}_n$, its cosets can be used to partition $\mathcal{G}_n$. Thus each error e belongs to a unique coset $e\mathcal{C}(\mathcal{S}) = \{ec|c \in \mathcal{C}(\mathcal{S})\}$. We state the following theorem without proof.[1]

Theorem 2.2. *Two errors $e_1, e_2 \in \mathcal{G}_n$ have the same error syndrome $l = l_1 \cdots l_{n-k}$ if and only if they belong to the same coset of $\mathcal{C}(\mathcal{S})$.*

Theorem 2.2 establishes that all elements of a coset of $\mathcal{C}(\mathcal{S})$ have the same error syndrome, and different cosets have different syndrome values. Thus each coset can be labeled by a unique syndrome value and so the number of cosets of $\mathcal{C}(\mathcal{S})$ cannot exceed the total number of syndrome values 2^{n-k}. It is possible to show[1] that the number of cosets of $\mathcal{C}(\mathcal{S})$ cannot be less than

64

2^{n-k}. Putting these two results together implies that there are exactly 2^{n-k} cosets of $\mathcal{C}(\mathcal{S})$ and that there is a $1-1$ correspondence between the set of syndrome values and the collection of cosets of the centralizer $\mathcal{C}(\mathcal{S})$ in $\mathcal{G}_n$.

We have seen that a QECC introduces an encoding map ξ that sends (i) unencoded k-qubit states $|u\rangle \in H_2^k$ to n-qubit codewords $|c\rangle = \xi|u\rangle$; and (ii) unencoded k-qubit operators $\mathcal{O} \in \mathcal{G}_k$ to encoded n-qubit operators $\overline{\mathcal{O}} = \xi\mathcal{O}\xi^\dagger$ that map codewords to codewords: $\overline{\mathcal{O}}|c\rangle = \xi\mathcal{O}\xi^\dagger\xi|u\rangle = \xi|u'\rangle = |c'\rangle$, where $|u'\rangle$ is the image of $|u\rangle$ under $\mathcal{O}$ and $|c'\rangle$ is the encoded image of $|u'\rangle$. In particular, the unencoded Pauli operators σ_x^i and σ_z^i are mapped to $X_i = \xi\sigma_x^i\xi^\dagger$ and $Z_i = \xi\sigma_z^i\xi^\dagger$, where $i = 1,\ldots,k$ labels the qubits. We will refer to the X_i, Z_i, and $Y_i = iZ_iX_i$ as the encoded Pauli operators. It follows from the commutation relations for the unencoded Pauli operators that

$$[X_i, X_j] = [Z_i, Z_j] = 0 \qquad (19)$$
$$[X_i, Z_j] = 0 \quad (i \neq j) \qquad (20)$$
$$\{X_i, Z_i\} = 0. \qquad (21)$$

Since unencoded operators $\mathcal{O} \in \mathcal{G}_k$, we can repeat the arguments leading from eq. (16) to eq. (17) to write

$$\mathcal{O} = i^\lambda \sigma_x(a)\sigma_z(b), \qquad (22)$$

where $a = a_1 \cdots a_k$ and $b = b_1 \cdots b_k$ are bit strings of length k and $\lambda = 0,1,2,3$. Encoding sends $\mathcal{O}$ to

$$\overline{\mathcal{O}} = \xi\left[i^\lambda \sigma_x(a)\sigma_z(b)\right]\xi^\dagger$$
$$= i^\lambda X(a)Z(b), \qquad (23)$$

where $X(a) = (X_1)^{a_1}\cdots(X_k)^{a_k}$ and $Z(b) = (Z_1)^{b_1}\cdots(Z_k)^{b_k}$. It is possible to show[1] that the encoding map ξ can always be chosen so that the encoded Pauli operators commute with the generators of the stabilizer $\mathcal{S}$ so that $X_i, Z_i \in \mathcal{C}(\mathcal{S})$ with $i = 1,\ldots k$. Furthermore, it can be shown[1] that $\mathcal{C}(\mathcal{S})$ is generated by the encoded Pauli operators $\{X_i, Z_i | i = 1,\ldots,k\}$ and the generators $\{g_j | j = 1,\ldots,n-k\}$. It follows from this that the cosets $\{X_i\mathcal{S}, Z_i\mathcal{S} : i = 1,\ldots,k\}$ generate the quotient group $\mathcal{C}(\mathcal{S})/\mathcal{S}$ and that encoded operations $\overline{\mathcal{O}}$ are in $1-1$ correspondence with the elements of the quotient group $\mathcal{C}(\mathcal{S})/\mathcal{S}$: $\overline{\mathcal{O}} \leftrightarrow \overline{\mathcal{O}}\mathcal{S}$.

As an example of a quantum stabilizer code we examine the class of QECCs known as Calderbank-Shor-Steane (CSS) codes.[9,10] As we shall see in Sec. 3, this class of quantum codes has a structure that makes them convenient for use in fault-tolerant quantum computing. CSS codes are constructed from two classical binary codes C and C' that have the following properties:

(1) C and C' are (n, k, d) and (n, k', d') codes, respectively;
(2) $C' \subset C$; and
(3) C and $C'_\perp$ are both t-error correcting codes.

The code construction first partitions C into the cosets of C':

$$C = C' \cup (c_1 + C') \cup \cdots \cup (c_N + C'),$$

where $c_1, \ldots, c_N \in C$, and N is the number of cosets of C' in C. From Lagrange's theorem, $N = 2^k / 2^{k'}$. Next $N = 2^{k-k'}$ basis codewords are defined by identifying each one with a coset of C': $|\bar{c}_i\rangle \leftrightarrow c_i + C'$. Specifically,

$$|\bar{c}_i\rangle = \frac{1}{\sqrt{2^{k'}}} \sum_{c' \in C'} |c_i + c'\rangle \quad (i = 1, \ldots, N). \tag{24}$$

Note that if e and f are codewords in C that belong to the same coset of C', they define the same basis codeword $|\bar{e}\rangle = |\bar{f}\rangle$. This follows since, if e and f belong to the same coset, there exists an $h \in C'$ such that $e = f + h$, and from eq. (24):

$$|\bar{e}\rangle = \frac{1}{\sqrt{2^{k'}}} \sum_{c' \in C'} |e + c'\rangle$$

$$= \frac{1}{\sqrt{2^{k'}}} \sum_{c' \in C'} |f + h + c'\rangle. \tag{25}$$

Since C' is a subspace of C, it is closed under addition and so $h' = h + c' \in C'$. Note that as c' ranges over C', so does h'. Thus eq. (25) can be rewritten as:

$$|\bar{e}\rangle = \frac{1}{\sqrt{2^{k'}}} \sum_{h' \in C'} |f + h'\rangle$$

$$= |\bar{f}\rangle. \tag{26}$$

The code space C_q is the subspace of H_2^n spanned by the $|\bar{c}_i\rangle$ and so is $N = 2^{k-k'}$ dimensional. Thus the CSS construction causes $k - k'$ qubits to be encoded into n qubits. It is possible to show that a CSS code can correct as many errors as do C and $C'_\perp$. The generators for a CSS code are constructed from the parity check matrices $H(C)$ and $H(C'_\perp)$. It can be shown that $C'_\perp$ maps $n - k'$ bits into n bits and so $H(C'_\perp)$ has k' rows. We construct $n - k$ generators by identifying a generator with each row of

$H(C)$ and replacing each 0 with the identity operator I and each 1 with σ_x^i, where i is the column label that now labels qubit i. The remaining k' generators are identified with the rows of $H(C'_\perp)$, only now $1 \to \sigma_z^i$. Thus

$$g_i = \left(\sigma_x^1\right)^{H_{i,1}(C)} \cdots \left(\sigma_x^n\right)^{H_{i,n}(C)} \qquad (i = 1, \ldots, n - k)$$
$$g_j = \left(\sigma_z^1\right)^{H_{j,1}(C'_\perp)} \cdots \left(\sigma_z^n\right)^{H_{j,n}(C'_\perp)} \qquad (j = 1, \ldots, k'). \qquad (27)$$

The following example constructs a CSS code that encodes 1 qubit into 7 qubits and corrects all single-qubit errors.

Example 2.1. For the $[7, 1, 3]$ CSS code, C is the $(7, 4, 3)$ binary Hamming code and C' the $(7, 3, 4)$ binary simplex code. These codes are presented in Ref. 2 where it is shown that $C'_\perp$ is the $(7, 4, 3)$ binary Hamming code. Thus $C = C'_\perp$ and so $H(C) = H(C'_\perp)$. We see that $n = 7$ and $k - k' = 4 - 3 = 1$. Thus the resulting CSS code will encode 1 qubit in 7 as promised above. The code distance d will be determined below. The parity check matrix $H(C)$ for the $(7, 4, 3)$ Hamming code is the matrix whose columns consist of all non-zero binary vectors of length 3:

$$H(C) = \begin{pmatrix} 0\,0\,0\,1\,1\,1\,1 \\ 0\,1\,1\,0\,0\,1\,1 \\ 1\,0\,1\,0\,1\,0\,1 \end{pmatrix}. \qquad (28)$$

The codewords for C are:[1,2]

$$\begin{array}{llll} 000\,0000 & 011\,0011 & 101\,0101 & 110\,0110 \\ 000\,1111 & 011\,1100 & 101\,1010 & 110\,1001 \\ 111\,1111 & 100\,1100 & 010\,1010 & 001\,1001 \\ 111\,0000 & 100\,0011 & 010\,0101 & 001\,0110; \end{array}$$

and those of C' are the even weight codewords in C which are seen to be the first eight codewords of C listed above. Thus the basis codewords for this CSS code are:

$$|\bar{0}\rangle = \frac{1}{\sqrt{2^3}} \{ |000\,0000\rangle + |011\,0011\rangle + |101\,0101\rangle + |110\,0110\rangle$$
$$+ |000\,1111\rangle + |011\,1100\rangle + |101\,1010\rangle + |110\,1001\rangle \}$$
$$|\bar{1}\rangle = \frac{1}{\sqrt{2^3}} \{ |111\,1111\rangle + |100\,1100\rangle + |010\,1010\rangle + |001\,1001\rangle$$
$$+ |111\,0000\rangle + |100\,0011\rangle + |010\,0101\rangle + |001\,0110\rangle \} .(29)$$

From $H(C) = H(C'_\perp)$ we have that:

$$
\begin{array}{ll}
g_1 = \sigma_x^4 \sigma_x^5 \sigma_x^6 \sigma_x^7 & ; \quad g_4 = \sigma_z^4 \sigma_z^5 \sigma_z^6 \sigma_z^7 \\
g_2 = \sigma_x^2 \sigma_x^3 \sigma_x^6 \sigma_x^7 & ; \quad g_5 = \sigma_z^2 \sigma_z^3 \sigma_z^6 \sigma_z^7 \\
g_3 = \sigma_x^1 \sigma_x^3 \sigma_x^5 \sigma_x^7 & ; \quad g_6 = \sigma_z^1 \sigma_z^3 \sigma_z^5 \sigma_z^7.
\end{array}
\tag{30}
$$

Direct calculation shows that the generators $g_1, \ldots, g_6$ fix the basis code-words $|\bar{0}\rangle$ and $|\bar{1}\rangle$. By definition, the code distance d for this code is the weight of the smallest error E that violates eq. (6). It is straightforward, though tedious, to check that all 1- and 2-qubit errors satisfy this equation, while the 3-qubit error $\sigma_x^1 \sigma_x^2 \sigma_x^3$ does not. In fact, $\langle \bar{0} | \sigma_x^1 \sigma_x^2 \sigma_x^3 | \bar{1} \rangle = 1 \neq 0$. Thus the code distance is $d = 3$ and the result of this construction is the $[7, 1, 3]$ CSS code. As we saw earlier, a code that corrects t errors must have distance $d = 2t + 1$. Thus our $[7, 1, 3]$ CSS code corrects all single-qubit errors. Finally, choosing $X = \sigma_x^1 \sigma_x^2 \sigma_x^3$ and $Z = \sigma_z^1 \sigma_z^2 \sigma_z^3$, direct calculation shows that: (i) $X|\bar{i}\rangle = |\overline{i \oplus 1}\rangle$ and $Z|\bar{i}\rangle = (-1)^i |\bar{i}\rangle$; (ii) $\{X, Z\} = 0$; and (iii) X and Z commute with all generators $g_1, \ldots, g_6$. Thus X and Z are the encoded Pauli operators.

2.4. *Concatenated Quantum Error Correcting Codes*

We have seen that an $[n_1, k_1, d_1]$ QECC uses entangled states of n_1 qubits to protect k_1 qubits of data. If error recovery can be applied perfectly, encoding will reduce the effective error probability for the data from $\mathcal{O}(p)$ to $\mathcal{O}(p^2)$. The question arises: if using an $[n_1, k_1, d_1]$ code improves the reliability of the k_1 data qubits, can further improvement be obtained by adding in another layer of encoding? Specifically, would splitting up the n_1 qubits into blocks of k_2 qubits and encoding each block using an $[n_2, k_2, d_2]$ code further reduce the effective error probability on the k_1 data qubits? As will be seen when discussing the accuracy threshold theorem in Sec. 4, under appropriate circumstances the answer is yes, and this layering in of entanglement by recursively encoding blocks of qubits is the essence of what concatenated QECCs do.[11,12] These codes will play an essential role in establishing the accuracy threshold theorem. This section explains how quantum stabilizer codes can be used to construct concatenated QECCs. The discussion will be limited to two layers of concatenation, though from what is said, it should be clear how to add in further layers. We will also restrict ourselves to the case where the second QECC encodes a single qubit. For a discussion of how to use a multi-qubit QECC to encode the second layer of entanglement, see Ref. 1.

To begin our construction, let the first layer of entanglement be introduced using an $[n_1, k_1, d_1]$ quantum stabilizer code $\mathcal{C}_1$ with generators $G_1 = \{g_i^1 \,|\, i = 1, \ldots, n_1 - k_1\}$, and the second layer introduced using an $[n_2, 1, d_2]$ code $\mathcal{C}_2$ with generators $\{g_j^2 \,|\, j = 1, \ldots, n_2 - 1\}$. The concatenated code $\mathcal{C}$ is to map k_1 qubits into $n = n_1 n_2$ qubits, with code construction parsing the n qubits into n_1 blocks $B(b)$ with $b = 1, \ldots, n_1$, each containing n_2 qubits. The generators of $\mathcal{C}$ are constructed as follows. (1) To each block $B(b)$ we associate a copy of the generators of $\mathcal{C}_2$: $G_2(b) = \{g_j^2(b) \,|\, j = 1, \ldots, n_2 - 1\}$. Each generator $g_j^2(b)$ is a block operator that only acts on qubits belonging to $B(b)$. With $n_2 - 1$ generators coming from each of the n_1 blocks, this step introduces a total of $n_1(n_2 - 1)$ generators. (2) Finally, we complete the set of generators of $\mathcal{C}$ by adding in the encoded images of the generators of $\mathcal{C}_1$. Specifically, if ξ_2 is the encoding operator for $\mathcal{C}_2$ and g_i^1 is a generator of $\mathcal{C}_1$, then $\overline{g_i^1} = \xi_2 g_i^1 \xi_2^\dagger$ is included as a generator of $\mathcal{C}$. This adds $n_1 - k_1$ generators to $\mathcal{C}$ for a total of $n_1(n_2 - 1) + n_1 - k_1 = n_1 n_2 - k_1$ generators. Since an $[n_1 n_2, k, d]$ code will have $n_1 n_2 - k$ generators, we see that the above construction produces an $[n_1 n_2, k_1, d]$ code, as desired. A general formula for the code distance d is not known, though the lower bound $d \geq d_1 d_2$ has been established.[1]

Example 2.2. Here we construct the Shor [9,1,3] code[13] by concatenating two [3,1,1] codes. The generators for the codes $\mathcal{C}_1$ and $\mathcal{C}_2$ are, respectively,

$$g_1^1 = \sigma_x^1 \sigma_x^2 \quad ; \quad g_2^1 = \sigma_x^1 \sigma_x^3 \tag{31}$$

and

$$g_1^2 = \sigma_z^1 \sigma_z^2 \quad ; \quad g_2^2 = \sigma_z^1 \sigma_z^3. \tag{32}$$

Choosing the unencoded computational basis states to be eigenstates of σ_z, $\sigma_z |i\rangle = (-1)^i |i\rangle$, the basis codewords $|\bar{i}\rangle_1$ for $\mathcal{C}_1$ are

$$
\begin{aligned}
|\bar{0}\rangle_1 &= \xi_1 |0\rangle \\
&= \frac{1}{\sqrt{2^3}} \left[|0\rangle + |1\rangle \right] \otimes \left[|0\rangle + |1\rangle \right] \otimes \left[|0\rangle + |1\rangle \right]; \\
|\bar{1}\rangle_1 &= \xi_1 |1\rangle \\
&= \frac{1}{\sqrt{2^3}} \left[|0\rangle - |1\rangle \right] \otimes \left[|0\rangle - |1\rangle \right] \otimes \left[|0\rangle - |1\rangle \right],
\end{aligned}
\tag{33}
$$

and for C_2 are

$$|\bar{0}\rangle_2 = \xi_2|0\rangle$$
$$= |0\rangle \otimes |0\rangle \otimes |0\rangle \equiv |000\rangle;$$
$$|\bar{1}\rangle_2 = \xi_2|1\rangle$$
$$= |1\rangle \otimes |1\rangle \otimes |1\rangle \equiv |111\rangle. \tag{34}$$

A subscript $i = 1, 2$ on an encoded ket indicates that it belongs to the code C_i. From eqs. (31) and (33) it is clear that C_1 maps $\sigma_x \rightarrow X = \sigma_z^1 \sigma_z^2 \sigma_z^3$ and $\sigma_z \rightarrow Z = \sigma_x^1 \sigma_x^2 \sigma_x^3$; and from eqs. (32) and (34) that C_2 maps $\sigma_x \rightarrow X = \sigma_x^1 \sigma_x^2 \sigma_x^3$ and $\sigma_z \rightarrow Z = \sigma_z^1 \sigma_z^2 \sigma_z^3$.

We now construct the generators of the concatenated code C. This code uses nine qubits that are parsed into three blocks, each containing three qubits. Qubits 1-3 belong to block 1; 4-6 to block 2; and 7-9 to block 3. To each block we associate a copy of the generators of C_2. This gives the following six generators:

$$g_1 = \sigma_z^1 \sigma_z^2$$
$$g_2 = \sigma_z^1 \sigma_z^3$$
$$g_3 = \sigma_z^4 \sigma_z^5$$
$$g_4 = \sigma_z^4 \sigma_z^6$$
$$g_5 = \sigma_z^7 \sigma_z^8$$
$$g_6 = \sigma_z^7 \sigma_z^9. \tag{35}$$

The remaining two generators are the encoded versions of g_1^1 and g_2^1. Here the qubit label i in eq. (31) becomes a block label, and each σ_x^i is replaced with $X_i = \sigma_x^{a_i} \sigma_x^{b_i} \sigma_x^{c_i}$. The superscripts a_i, b_i, and c_i are determined by which block is being encoded. Thus,

$$g_7 = X_1 X_2$$
$$= \left(\sigma_x^1 \sigma_x^2 \sigma_x^3\right) \left(\sigma_x^4 \sigma_x^5 \sigma_x^6\right);$$
$$g_8 = X_1 X_3$$
$$= \left(\sigma_x^1 \sigma_x^2 \sigma_x^3\right) \left(\sigma_x^7 \sigma_x^8 \sigma_x^9\right). \tag{36}$$

Eqs. (35) and (36) are the generators for the concatenated code C. It is straightforward to show that all one- and two-qubit errors satisfy eq. (6), while $e = \sigma_x^1 \sigma_x^2 \sigma_x^3$ violates it. Thus C has distance $d = 3$ and is a [9,1,3] QECC. In fact, it is the Shor [9,1,3] code.[13] Note that $d = 3$ is consistent with the lower bound $d \geq d_1 d_2 = 1$. The basis codewords for $\mathcal{C}$ are

$$
\begin{aligned}
|\bar{0}\rangle &= \xi_2 \xi_1 |0\rangle \\
&= \frac{1}{\sqrt{2^3}} [|000\rangle + |111\rangle] \otimes [|000\rangle + |111\rangle] \otimes [|000\rangle + |111\rangle] ; \\
|\bar{1}\rangle &= \xi_2 \xi_1 |1\rangle \\
&= \frac{1}{\sqrt{2^3}} [|000\rangle - |111\rangle] \otimes [|000\rangle - |111\rangle] \otimes [|000\rangle - |111\rangle] , \quad (37)
\end{aligned}
$$

and the encoded Pauli operators $X = (\xi_2 \xi_1) \sigma_x (\xi_2 \xi_1)^\dagger$ and $Z = (\xi_2 \xi_1) \sigma_z (\xi_2 \xi_1)^\dagger$ are

$$
\begin{aligned}
X &= \sigma_z^1 \sigma_z^2 \sigma_z^3 \sigma_z^4 \sigma_z^5 \sigma_z^6 \sigma_z^7 \sigma_z^8 \sigma_z^9 \\
Z &= \sigma_x^1 \sigma_x^2 \sigma_x^3 \sigma_x^4 \sigma_x^5 \sigma_x^6 \sigma_x^7 \sigma_x^8 \sigma_x^9 .
\end{aligned} \quad (38)
$$

It is straightforward to check that $Z|\bar{i}\rangle = (-1)^i |\bar{i}\rangle$ and $X|\bar{i}\rangle = |\overline{i \oplus 1}\rangle$, where $\oplus$ is addition modulo 2.

3. Fault-Tolerant Quantum Computing

Having discussed quantum error correcting codes, we move on to a discussion of fault-tolerant quantum computing (FTQC). The discussion will take place in the context of quantum stabilizer codes. We begin by introducing the idea of fault-tolerance and making basic definitions in Sec. 3.1, then describe a procedure for doing fault-tolerant quantum error correction in Sec. 3.2. The theory of FTQC is then taken up in Sec. 3.3. The presentation breaks up into three parts. In the first part (Secs. 3.3.1–3.3.4), we introduce the Clifford group and describe how quantum gates in this group can be implemented fault-tolerantly on data encoded using an $[n, 1, d]$ quantum stabilizer code. The second part (Sec. 3.3.5) describes an extension of this approach to quantum stabilizer codes that encode multiple qubits. Finally, Sec. 3.3.6 describes how a fault-tolerant Toffoli gate can be constructed for any quantum stabilizer code. Since the Toffoli and Clifford group gates together form a universal set of quantum gates, this three-part development establishes how to do fault-tolerant universal quantum computing on data encoded using any quantum stabilizer code.

3.1. Motivation

Noise and imperfect quantum gates hamper the operation of a quantum computer in both direct and indirect ways. Directly in that they cause errors to appear in the computational data, and indirectly as computation usually requires the application of operations that are controlled by the corrupted data. The corrupted data cause wrong operations to be applied, and so new errors are generated from existing errors. To illustrate how such error proliferation can occur, consider two qubits, the first in the state $|1\rangle$ and the second in the generic state $a|0\rangle + b|1\rangle$. The two-qubit state is then $|1\rangle \otimes [a|0\rangle + b|1\rangle]$. Suppose noise induces a bit-flip error on the first qubit ($E = \sigma_x^1$) with probability p. This switches the two-qubit state to $|\psi_{in}\rangle = |0\rangle \otimes [a|0\rangle + b|1\rangle]$. Now suppose a CNOT gate is applied with the first (second) qubit as the control (target).[14] The output state is then

$$\begin{aligned}
|\psi_{out}\rangle &= U_{CNOT}|\psi_{in}\rangle \\
&= |0\rangle \otimes [a|0\rangle + b|1\rangle] \\
&= \sigma_x^1 \sigma_x^2 |\psi_c\rangle,
\end{aligned} \qquad (39)$$

where $|\psi_c\rangle = |1\rangle \otimes [a|1\rangle + b|0\rangle]$ is the state that would have resulted from the CNOT gate if the original error on qubit 1 had not occurred. We see from eq. (39) that the original bit-flip error σ_x^1 on the control qubit has generated a second bit-flip error σ_x^2 on the target qubit even though the CNOT gate executed *correctly*. Thus two errors have appeared in $|\psi_{out}\rangle$ with probability p! If the two qubits had simply been sitting in storage, and if errors occurred independently, two errors would occur with probability p^2. However, because a quantum computer actively manipulates its data, errors can multiply, and so it is possible for two or more errors to appear in the data with probability p. Similarly, it is possible to show that an error-free application of a CNOT gate that uses qubit 1 as the control will cause a phase error present on the target qubit to spread to a second phase error on the control qubit $\sigma_z^2 \to \sigma_z^1 \sigma_z^2$. It is the goal of *fault-tolerant design* to provide quantum circuits that control the manner in which errors spread during (i) encoding/decoding, (ii) error correction and recovery, and (iii) encoded operations. As will be seen in Sec. 4, when a sufficiently layered concatenated QECC is used to protect the data, and a sufficiently reliable universal set of quantum gates is available, these fault-tolerant quantum circuits will allow a quantum computation of arbitrary duration to be done with arbitrarily small error probability.

We are now ready to define a fault-tolerant quantum operation.

Definition 3.1. Suppose we are given a quantum register composed of several blocks of qubits and a QECC with which to encode data into each block. An operation is said to be fault-tolerant if the occurrence of a single gate error or storage error during the course of the operation produces no more than one error in each code block.

For QECCs that correct more than one error, this restriction to only one error per block might seem overly consrevative. Perhaps so, though this "one error per block" restriction simplifies the application of Definition 3.1 to specific operations. As will be seen in the remainder of this section, even with this tougher condition, FTQC will prove possible.

We will use the following error model in the remainder of this Chapter. We assume that the environment of the quantum register: (i) causes errors on different qubits to occur independently; (ii) single-qubit errors σ_x^i, σ_y^i, and σ_z^i are equally likely; (iii) the single-qubit error probability p_i is the same for all qubits $p_i = p$; and (iv) gate errors only affect the qubits acted on by the gate. Clearly, other error models are possible. The one presented here has the merit of being simple to analyze while still being sufficiently realistic to be physically interesting.

For later applications, it proves convenient to make two simple restatements of Definition 3.1.

(1) An operation is fault-tolerant if, to first order in p (viz. $\mathcal{O}(p)$), the operation generates no more than one error per block.

(2) An operation fails to be fault-tolerant if, to $\mathcal{O}(p)$, it can generate more than one error per code block.

The following definition introduces a transversal operation, which is then shown to be fault-tolerant.

Definition 3.2. A transversal operation satisfies one of the following two conditions. (1) It only applies one-qubit gates to the qubits in a code block. (2) It only interacts the ith qubit in one code block with the ith qubit in a different code block or block of ancilla qubits.

It is straight-forward to check that transversal operations are fault-tolerant. For one-qubit transversal operations, fault-tolerance is automatic since, if only one gate error or storage error occurs, only one qubit is affected. Thus only one error appears in the code block and Definition 3.1 is satisfied. For two-qubit transversal operations, suppose the operation that interacts the ith qubits in the two different blocks fails. Since they are in separate blocks, at most one error appears in each block and so Definition 3.1 is again satisfied. Now consider a storage error. Suppose one appears on qubit i in the first block before it is interacted with the ith qubit in the second block. Upon interaction, this error cannot propogate beyond these two qubits since only they take part in the interaction. Since they are in separate blocks, at most one error appears in each block and so the definition of fault-tolerance is satisfied. Finally, if the storage error occurs after the two i^{th} qubits have been interacted, only one code block picks up an error and so Definition 3.1 is satisfied in this case as well. Transversal operations are thus seen to be fault-tolerant.

3.2. *Fault-Tolerant Quantum Error Correction*

Here we present a fault-tolerant protocol for quantum error correction. This protocol implements the error correction scenario described in Sec. 2.2. Due to space limitations, a detailed presentation is not possible. The reader can find a complete discussion in Ref. 1.

3.2.1. *Syndrome Extraction*

Recall that for an error $E \in \mathcal{G}_n$ and an $[n, k, d]$ quantum stabilizer code, the syndrome for E is the bit string $S(E) = l_1 \cdots l_{n-k}$, where the syndrome bit l_i is determined by whether E commutes or anticommutes with the generator g_i:

$$g_i E = (-1)^{l_i} E g_i \quad (i = 1, \ldots, n - k). \tag{40}$$

It follows that the corrupted codeword $E|c\rangle$ is a simultaneous eigenstate of the generators $\{g_1, \ldots, g_{n-k}\}$:

$$g_i \{E|c\rangle\} = (-1)^{l_i} \{E|c\rangle\} \quad (i = 1, \ldots, n - k). \tag{41}$$

Measuring g_i when the quantum register is in the state $E|c\rangle$ returns the eigenvalue $(-1)^{l_i}$, and so determines l_i. The syndrome $S(E) = l_1 \cdots l_{n-k}$ is then found by measuring all $n - k$ generators.

In syndrome extraction ancilla qubits are interacted with a code block in such a way that the syndrome $S(E)$ is encoded into the state of the ancilla. An appropriate measurement of the ancilla then yields $S(E)$. We now describe how this can be done using a quantum circuit that only applies transversal operations so that the circuit itself is fault-tolerant. We shall see, however, that fault-tolerant preparation of the ancilla state and verification of the measured syndrome value require a separate analysis. These tasks are taken up in Secs. 3.2.2 and 3.2.3, respectively.

As shown in Theorem 2.2, $S(E)$ determines the coset of the centralizer $\mathcal{C}(\mathcal{S})$ that contains the error E. Let E_* denote a smallest weight error in this coset. For our error model, E_* corresponds to a most probable error for this coset. If more than one error has smallest weight, one arbitrarily chooses one of these smallest weight errors to be E_*. We then use $E_*^\dagger = E_*^{-1}$ to implement the error recovery operation: $|\psi_{out}\rangle = E_*^{-1} E|c\rangle$. Since E_* is a most probable error, the recovery operation succeeds with high probability. If, however, $E \neq E_*$, then $|\psi_{out}\rangle \neq |c\rangle$, and the recovery operation fails to correct the error.

Recall that each generator can be written as $g_i = i^{\lambda_i} \sigma_x(a_i)\sigma_z(b_i)$, where $a_i = a_{i,1} \cdots a_{i,n}$ and $b_i = b_{i,1} \cdots b_{i,n}$ are bit strings of length n. Let $\mathbf{H}_i$ denote the following transversal operation:

$$\mathbf{H}_i = \prod_{j=1}^{n} (H_j)^{a_{i,j}}, \tag{42}$$

where H_j is the Hadamard gate applied to qubit j. Finally, let $U_{CNOT}(a_i + b_i)$ denote the following transversal application of CNOT gates from the code block to the ancilla block:

$$U_{CNOT}(a_i + b_i) = \prod_{j=1}^{n} (U_{CNOT})^{a_{i,j}+b_{i,j}}. \tag{43}$$

As we shall see below, this operation will be used to transfer the syndrome bit l_i to the ancilla state.

As explained above, to determine the syndrome bit l_i, we must measure the generator g_i when the quantum register is in the corrupted state $E|c\rangle$. Since $g_i, E \in \mathcal{G}_n$, we can write (suppressing factors of i^λ): $g_i = \sigma_x(a_i)\sigma_z(b_i)$ and $E = \sigma_x(e_x)\sigma_z(e_z)$. In the following it is assumed that g_i contains no factors of σ_y. This restriction is not essential and can be relaxed,[1] though making it simplifies the following analysis. Inserting these forms into eq. (40), one can show that $l_i = e_z \cdot a_i \oplus e_x \cdot b_i$, where the scalar product has the usual form, although addition is modulo 2. Let g_i have weight w_i. To determine its associated syndrome bit l_i we introduce w_i ancilla bits. The first step of the syndrome extraction protocol is to apply $\mathbf{H}_i$ to the quantum register. Since this is a transversal operation, this step is fault-tolerant. The register state is then $\mathbf{H}_i E|c\rangle = \overline{E}|\overline{c}\rangle$, where $\overline{E} = \mathbf{H}_i E \mathbf{H}_i^\dagger$ and $|\overline{c}\rangle = \mathbf{H}_i|c\rangle$. It is simple to show that this state is an eigenstate of $\overline{g}_i = \mathbf{H}_i g_i \mathbf{H}_i^\dagger = \sigma_z(a_i + b_i)$ with eigenvalue $(-1)^{l_i}$. Because the transformed generator $\overline{g}_i$ only contains factors of σ_z and its eigenvalue also determines l_i, it proves more convenient to work with than g_i. The ancilla qubits are assumed to have been prepared in the Shor state $|A\rangle$:

$$|A\rangle = 2^{-(w_i-1)/2} \sum_{A_e} |A_e\rangle, \tag{44}$$

where the sum is over all even-parity bit strings A_e of length w_i. Fault-tolerant preparation of the Shor state is described in Sec. 3.2.2. Next, the protocol applies $U_{CNOT}(a_i + b_i)$ to the quantum register. As noted earlier, this operation is transversal and so is fault-tolerant. One can show[1] that this leaves the register and ancilla block in the state:

$$U_{CNOT}(a_i + b_i)\overline{E}|\overline{c}\rangle \otimes |A\rangle = \overline{E}|\overline{c}\rangle \otimes \sum_{A_e} |A_e \oplus l_i\rangle. \tag{45}$$

We see that the label of each ancilla state in the sum contains the syndrome bit l_i. Applying $\mathbf{H}_i$ again leaves the quantum register and ancilla block in the state

$$\mathbf{H}_i U_{CNOT}\mathbf{H}_i E|c\rangle \otimes |A\rangle = E|c\rangle \otimes \sum_{A_e} |A'_e + l_i\rangle. \tag{46}$$

Because this state is non-entangled, measurement of the ancilla block will not cause any superposition of states in the code block to collapse. If we measure the ancilla block, the result will be

$$m_i = (-1)^{wt(A'_e)\oplus l_i}, \tag{47}$$

for some A'_e. Note that since A'_e is an even weight bit string, $wt(A'_e) = 0$ modulo 2, and so $m_i = (-1)^{l_i}$ which determines l_i. Upon completion of the measurement the ancilla block is discarded. Repeating this procedure with a fresh ancilla block for each of the $n - k$ generators yields the error syndrome $S(E)$ and leaves the register in the state $E|c\rangle$. The error syndrome $S(E)$ determines the most probable error E_*, and thus the recovery operator $\mathbf{R} = E_*^\dagger$. Before going on to examine fault-tolerant procedures for Shor state preparation and syndrome verification, we illustrate syndrome extraction for the [5,1,3] quantum stabilizer code.

Example 3.1. The generators for the [5,1,3] quantum stabilizer code are:[1]

$$g_1 = \sigma_x(10010)\sigma_z(01100)$$
$$g_2 = \sigma_x(01001)\sigma_z(00110)$$
$$g_3 = \sigma_x(10100)\sigma_z(00011)$$
$$g_4 = \sigma_x(01010)\sigma_z(10001), \tag{48}$$

which allows the bit strings a_i and b_i to be read off for each generator g_i, and determines the weight $w_i = 4$ for $i = 1, \ldots, 4$. It follows from eq. (42) that

$$\mathbf{H}_1 = H_1 H_4$$
$$\mathbf{H}_2 = H_2 H_5$$
$$\mathbf{H}_3 = H_1 H_3$$
$$\mathbf{H}_4 = H_2 H_4, \tag{49}$$

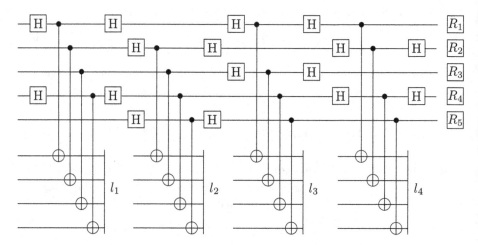

Fig. 1.   Quantum circuit for syndrome extraction for the [5,1,3] quantum stabilizer code. The upper five lines correspond to the code block and each set of lower four lines to an ancilla block. Qubit indicies increase in going from top to bottom in each block. The measured syndrome value $S(E) = l_1 l_2 l_3 l_4$ determines the recovery operation $\mathbf{R} = R_1 R_2 R_3 R_4 R_5$.

and so

$$\bar{g}_1 = \sigma_z(11110)$$
$$\bar{g}_2 = \sigma_z(01111)$$
$$\bar{g}_3 = \sigma_z(10111)$$
$$\bar{g}_4 = \sigma_z(11011). \tag{50}$$

Eqs. (46) and (50) determine the quantum circuit for syndrome extraction that appears in Fig. 1. This circuit encodes the syndrome bits into four ancilla blocks where they can be measured, and the result used to determine the transversal (and hence fault-tolerant) recovery operation $\mathbf{R} = R_1 R_2 R_3 R_4 R_5$. The syndrome bit l_i is transferred to the ith ancilla block in accordance with eq. (46). Note that Plenio et al.[15] have shown that there are advantages to rearranging the sequence of CNOT gates in Fig. 1, though space limitations do not allow us to describe that work here.

3.2.2. Shor State Verification

As noted in Sec. 3.2.1, although the syndrome extraction circuit only uses transversal operations and so is fault-tolerant, we do not yet have a fault-

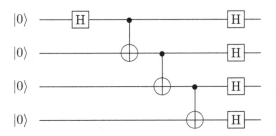

Fig. 2. Quantum circuit to prepare the Shor state $|A\rangle$ for $w_i = 4$. The first Hadamard gate and three CNOT gates place the ancilla block in the cat-state. The remaining four Hadamard gates transform this state into $|A\rangle$.

tolerant procedure for quantum error correction. The remaining difficulties stem from errors that may be present in the Shor state. As we shall see in Sec. 3.2.3, bit-flip errors in the Shor state can cause an error in the mesured syndrome value. Such an error will generate new errors in the data by causing the wrong recovery operation to be applied to the code block. A fault-tolerant procedure is thus needed to verify the measured syndrome value. Phase errors in the Shor state can also ruin fault-tolerance by spreading to the code block through the CNOT gates used during syndrome extraction. This subsection will deal with the difficulties arising from phase errors in the Shor state, while the following subsection will examine the case of bit-flip errors.

The ancilla block can be placed in the Shor state $|A\rangle$ using a two-step procedure.[16] First the block is prepared in the cat-state

$$|\psi_{cat}\rangle = \frac{1}{\sqrt{2}} \left\{ |0\cdots0\rangle + |1\cdots1\rangle \right\}, \qquad (51)$$

which is a uniform superposition of states in which the ancilla qubits are all in the $|0\rangle$ CB state or all in the $|1\rangle$ CB state. The procedure is completed by applying a Hadamard gate to each ancilla qubit. Direct calculation shows that the Hadamard gates transform $|\psi_{cat}\rangle \to |A\rangle$.

A quantum circuit to prepare $|A\rangle$ appears in Fig. 2. To make the discussion concrete we have set $w_i = 4$ as would be appropriate for the [5,1,3] code examined in Example 3.1. Generalizing to arbitrary w_i is straightforward. Initially all ancilla qubits are in the $|0\rangle$ CB state. The first Hadamard gate and three CNOT gates put the ancilla block in the cat-state and the final four Hadamard gates transform the cat-state into the Shor state $|A\rangle$.

Although the final four Hadamard gates are applied transversely and so operate fault-tolerantly, the same is not true of the protion of the circuit that creates the cat-state. Failure of one of the CNOT gates can lead to multiple errors in both the ancilla and code blocks. Specifically, suppose the second CNOT gate fails, causing a bit-flip error on its target qubit. Since this qubit becomes the control for the third CNOT gate, its bit-flip error spreads to the target qubit as shown in Sec. 3.1. Thus one gate error has produced two errors in the ancilla block with probability of $\mathcal{O}(p)$. But things get worse. After the final Hadamard gates are applied, the two bit flip errors become phase errors in the Shor state since $H\sigma_x = \sigma_z H$. As seen in Sec. 3.1, these two phase errors will propagate back into the code block via the CNOT gates used for syndrome extraction. Thus one gate error during cat-state preparation has produced two errors in both the code and ancilla blocks with probability of $\mathcal{O}(p)$. This procedure for Shor state preparation is thus not fault-tolerant. To recover fault-tolerance, the preparation procedure must be modified so that the probability for this outcome becomes of $\mathcal{O}(p^2)$.

To that end, following Shor,[16,17] notice that when the target for one of the CNOT gates used in cat-state preparation develops a bit-flip error, the error propagates to the target qubits of all subsequent CNOT gates in Fig. 2. In the cat-state, the first and last ancilla qubits always have even relative parity. A single CNOT gate error that causes a bit-flip error on its target qubit causes the relative parity of the first and last ancilla qubits to become odd. Figure 3 gives a modified quantum circuit for Shor state preparation that takes advantage of this observation. A new ancilla qubit is introduced in the CB state $|0\rangle$ and it is made the target of CNOT gates from the first and fourth ancilla qubits. If these qubits have odd relative parity, the new ancilla qubit will wind up in the state $|1\rangle$. Measuring σ_z for the new qubit returns $s = 1$, telling us that an odd number of bit-flip errors occurred during cat-state preparation. In particular, to $\mathcal{O}(p)$, one bit-flip error occurred. In this case the ancilla block is discarded, a new ancilla block is introduced, and cat-state preparation is re-tried on the new ancilla block. If measurement of the new ancilla qubit yields $s = 0$, we know its state is $|0\rangle$, and that an even number of bit-flip errors has occurred. Specifically, to $\mathcal{O}(p)$, we know that no bit-flip errors occurred. In this case the final Hadamard gates can be applied, and we are assured that to $\mathcal{O}(p)$, the output Shor state will not contain phase errors. Said another way, this modified procedure insures that the probability that two phase errors appear in the Shor state is $\mathcal{O}(p^2)$, as desired.

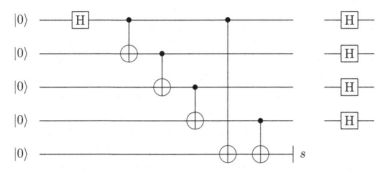

Fig. 3. Modified quantum circuit for Shor state preparation for $w_i = 4$. A new ancilla qubit is introduced to record the relative parity s of the first and fourth qubits. If measurement yields $s = 1$, the ancilla block is discarded, a new ancilla block is introduced, and the circuit is applied again. The final Hadamard gates are only applied if $s = 0$. See text for full discussion. Based on Fig. 11, J. Preskill, *Proc. R. Soc. Lond. A* **454** (1998) 385; ©Royal Society; used with permission.

3.2.3. *Syndrome Verification*

Bit-flip errors in the Shor state can cause the measured syndrome value to be in error. This is most easily seen from eq. (47). If an odd number of bit-flip errors occur, $wt(A'_e)$ will be an odd rather than an even integer, and the syndrome measurement will return $l_i \oplus 1$ instead of l_i. As pointed out in Sec. 3.2.2, bit-flip errors in the Shor state can appear during its preparation. Other possible sources are (i) a storage error that causes a bit-flip error on an ancilla qubit, or (ii) a faulty CNOT gate during syndrome extraction that introduces errors in both the code and ancilla blocks. In the latter case, if the error in the ancilla block is a bit-flip error, then the measured syndrome value will be incorrect, and if blindly accepted, will produce a second error in the code block through application of the wrong recovery operation to that block. Thus to $\mathcal{O}(p)$, one gate error produces two errors in the code block, and the error correction procedure fails to be fault-tolerant. The following procedure[15] insures that two errors appear in the code block only when two or more independent errors occur. We first state it, then explain how it leads to fault-tolerance.

(1) **Syndrome measurement returns $S = 0$:** Accept the measurement result as correct; no recovery operation applied to code block.
(2) **Syndrome measurement returns $S \neq 0$:** Remeasure syndrome using new ancilla block; accept the second syndrome result as correct; apply corresponding recovery operator to code block.

It can be shown[1] that the syndrome value $S = 0$ does not guarantee that the code block is error-free. In fact, to $\mathcal{O}(p)$, the code block can contain one error. By accepting $S = 0$ as the correct syndrome value and leaving the code block alone, the protocol does not introduce any new errors into the data. To $\mathcal{O}(p)$ then, at most an undetected error is present in the code block after error correction. The protocol is thus fault-tolerant for $S = 0$. If the syndrome measurement yields $S \neq 0$, at least one error has occurred, and to $\mathcal{O}(p)$, exactly one error has occurred. The above protocol requires that the syndrome be remeasured using a new ancilla block, and the new result be accepted as correct. This works since to $\mathcal{O}(p)$, for $S \neq 0$, one error already occurred during the first syndrome extraction, and so to first-order in p, no error can occur during the second syndrome extraction. To $\mathcal{O}(p)$, the second syndrome result will be correct. For the second syndrome result to be wrong, a new error would have to occur during the second syndrome extraction, making the overall process second-order in p. The recovery operator corresponding to the second syndrome result can be applied to the code block, and we are assured recovery succeeds to first-order in p. The protocol is thus fault-tolerant for $S \neq 0$ as well.

Ref. 15 has extended this protocol so that it only fails when $t + 1$ or more errors occur during syndrome extraction. At most $t + 1$ syndrome measurements are required to determine the correct syndrome to $\mathcal{O}(p^t)$.

Steane has introduced an alternative approach for fault-tolerant quantum error correction. The reader is referred to Refs. 18,19 for a discussion of this approach.

3.3. Fault-Tolerant Universal Quantum Computing

A set of quantum gates $\mathcal{S}$ is said to be universal if it contains a finite number of elements, and an arbitrary unitary operation (viz. quantum computation) can be done with arbitrarily small error probability using a quantum circuit[14] that only contains gates in $\mathcal{S}$. Many universal sets of quantum gates are known, although we will only be interested in the set composed of the Hadamard gate H, the phase gate P, the CNOT gate U_{CNOT}, and the Toffoli gate U_T. Shor[16] and Kitaev[20] have shown that this set is universal. Let $|i\rangle$, $|ij\rangle$, and $|ijk\rangle$ be the one-, two-, and three-qubit CB states, respectively, with $i, j, k = 0, 1$. The action of these gates is determined by their action on the CB states:

$$H|j\rangle = \frac{1}{\sqrt{2}}\left[|0\rangle + (-1)^{j}|1\rangle\right];$$
$$P|j\rangle = e^{ij\pi/2}|j\rangle;$$
$$U_{CNOT}|ij\rangle = |i(j \oplus i)\rangle;$$
$$U_{T}|ijk\rangle = |ij(k \oplus ij)\rangle, \tag{52}$$

where $\oplus$ is addition modulo 2. In this section we describe how all gates in this universal set can be applied fault-tolerantly on data encoded using any quantum stabilizer code. This then gives us the ability to do encoded universal fault-tolerant quantum computing.

3.3.1. *Clifford Group Gates*

Formally, an encoded quantum gate is a unitary operator U that maps codewords to codewords. As we now know, for a quantum stabilizer code $\mathcal{C}_q$ with stabilizer $\mathcal{S}$, each codeword is fixed by every stabilizer element s: $s|c\rangle = |c\rangle$. It is easy to show that for all $|c\rangle \in \mathcal{C}_q$ and $s \in \mathcal{S}$, $U|c\rangle$ is fixed by $UsU^{\dagger}$:

$$\left(UsU^{\dagger}\right)\left(U|c\rangle\right) = U\,s|c\rangle = U|c\rangle. \tag{53}$$

Now if U fixes $\mathcal{S}$ under conjugation $\left(U\mathcal{S}U^{\dagger} = \mathcal{S}\right)$, then every element of $\mathcal{S}$ can be written as $UsU^{\dagger}$ for some $s \in \mathcal{S}$. Then for every codeword $|c\rangle$, it follows from eq. (53) that $U|c\rangle$ is a codeword since it is fixed by every element of $\mathcal{S}$. Thus U maps codewords to codewords whenever $U\mathcal{S}U^{\dagger} = \mathcal{S}$, and so is an encoded operation. Unitary operators that fix $\mathcal{S}$ under conjugation belong to a subgroup of the unitary group $U(n)$ known as the normalizer $N_{U}(\mathcal{S})$. It is possible to show[1] that the code centralizer $\mathcal{C}(\mathcal{S})$ is the normalizer $N_{\mathcal{G}}(\mathcal{S})$ of $\mathcal{S}$ in the Pauli group $\mathcal{G}_n$. Since $\mathcal{G}_n \subset U(n)$, it follows that $N_{\mathcal{G}}(\mathcal{S}) \subset N_{U}(\mathcal{S})$. For the remainder of Sec. 3 we will be solely interested in $N_{U}(\mathcal{S})$ and so we will suppress the subscript U: $N_{U}(\mathcal{S}) \to N(\mathcal{S})$.

Having indicated why the normalizer $N(\mathcal{S})$ is central to our discussion of encoded operations, we next introduce the Clifford group $N_{U}(\mathcal{G}_n)$, which is the normalizer of the Pauli group $\mathcal{G}_n$ in $U(n)$.[21-23] We again suppress the subscript U so that $N_{U}(\mathcal{G}_n) \to N(\mathcal{G}_n)$. We shall see below that the groups $N(\mathcal{S})$ and $N(\mathcal{G}_n)$ intersect, though one group need not be a subset of the other. For purposes of this section, the Clifford group $N(\mathcal{G}_n)$ is of interest because it is generated by a simple set of quantum gates: Hadamard H, phase P, and CNOT U_{CNOT}.[21,23] Because $N(\mathcal{G}_n)$ has such a simple generator set, we will focus on unitary operators $U \in N(\mathcal{G}_n) \cap N(\mathcal{S})$, which are thus encoded operations that can be constructed using Hadamard, phase, and CNOT gates. The examples we will consider in this subsection yield encoded operations that are transversal and so *fault-tolerant*.

We have just seen that a unitary operator U that fixes $\mathcal{S}$ under conjugation is an encoded operation. We saw in Sec. 2.3 that there is a one-to-one correspondence between encoded operations and the cosets of $\mathcal{C}(\mathcal{S})/\mathcal{S}$. It is possible to show[1] that a U that fixes $\mathcal{S}$ (under conjugation) maps $\mathcal{C}(\mathcal{S})/\mathcal{S} \to \mathcal{C}(\mathcal{S})/\mathcal{S}$ and so maps encoded operations to encoded operations. One can also prove the following theorem.[1]

Theorem 3.1. *Let U be an element of $N(\mathcal{G}_n)$ which acts on $\mathcal{G}_n$ by conjugation: $f_U(g) = UgU^{\dagger}$ for all $g \in \mathcal{G}_n$. The map f_U is an automorphism of $\mathcal{G}_n$ (viz. f_U is an isomorphism from $\mathcal{G}_n \to \mathcal{G}_n$).*

Since each $U \in N(\mathcal{G}_n)$ is an automorphism of $\mathcal{G}_n$, we can determine the action of U on $\mathcal{G}_n$ once we know how the generators of $N(\mathcal{G}_n)$ act on the generators of $\mathcal{G}_n$. To that end, we now write out the action of the one-qubit Hadamard and phase gates on the generators of $\mathcal{G}_1$, and the two-qubit CNOT gate on the generators of $\mathcal{G}_2 = \mathcal{G}_1 \otimes \mathcal{G}_1$. The derivation of these actions is straightforward and will not be included here:

Hadamard gate H:

$$\sigma_x \to H\sigma_x H^{\dagger} = \sigma_z$$
$$\sigma_z \to H\sigma_z H^{\dagger} = \sigma_x$$
$$\sigma_y \to H\sigma_y H^{\dagger} = -\sigma_y; \tag{54}$$

Phase gate P:

$$\sigma_x \to P\sigma_x P^{\dagger} = \sigma_y$$
$$\sigma_z \to P\sigma_z P^{\dagger} = \sigma_z$$
$$\sigma_y \to P\sigma_y P^{\dagger} = -\sigma_x; \tag{55}$$

CNOT gate U_{CNOT}:

$$\sigma_x \otimes I \to U_{CNOT}\{\sigma_x \otimes I\}U_{CNOT}^{\dagger} = \sigma_x \otimes \sigma_x$$
$$I \otimes \sigma_x \to U_{CNOT}\{I \otimes \sigma_x\}U_{CNOT}^{\dagger} = I \otimes \sigma_x$$
$$\sigma_z \otimes I \to U_{CNOT}\{\sigma_z \otimes I\}U_{CNOT}^{\dagger} = \sigma_z \otimes I$$
$$I \otimes \sigma_z \to U_{CNOT}\{I \otimes \sigma_z\}U_{CNOT}^{\dagger} = \sigma_z \otimes \sigma_z. \tag{56}$$

Another useful two-qubit gate is the controlled-phase gate U_{CP}. Its action in the two-qubit CB states is: $U_{CP}|ij\rangle = (-1)^{i \cdot j}|ij\rangle$ with $i, j = 0, 1$. It is simple to show that $U_{CP} = (I \otimes H)U_{CNOT}(I \otimes H)$, and so is an element of the Clifford group $N(\mathcal{G}_n)$. It's action on the generators of $\mathcal{G}_2$ is:

Controlled-phase gate U_{CP}:

$$\sigma_x \otimes I \rightarrow U_{CP}\{\sigma_x \otimes I\}U_{CP}^\dagger = \sigma_x \otimes \sigma_z$$
$$I \otimes \sigma_x \rightarrow U_{CP}\{I \otimes \sigma_x\}U_{CP}^\dagger = \sigma_z \otimes \sigma_x$$
$$\sigma_z \otimes I \rightarrow U_{CP}\{\sigma_z \otimes I\}U_{CP}^\dagger = \sigma_z \otimes I$$
$$I \otimes \sigma_z \rightarrow U_{CP}\{I \otimes \sigma_z\}U_{CP}^\dagger = I \otimes \sigma_z. \tag{57}$$

As an application of the above discussion, we describe the effect of applying the generators of the Clifford group $N(\mathcal{G}_n)$ transversally to qubit blocks encoded using a special type of CSS code. For this class of CSS code, the approach of this subsection is sufficiently powerful to allow fault-tolerant quantum computation (when combined with the Toffoli gate construction of Sec. 3.3.7). Recall that a CSS code is constructed using a pair of classical binary codes C and C' with the following properties:

(1) C and C' are $[n, k, d]$ and $[n, k', d']$ codes, respectively;
(2) $C' \subset C$; and
(3) C and $C'_\perp$ are both t-error correcting codes.

Two further restrictions are now made. First, we require $C' = C_\perp$ so that condition 3 is automatically satisfied. Finally, we require that all codeword $c \in C$ have weights be divisible by 4: $wt(c) \equiv 0 \pmod 4$. Such codes are called doubly-even codes. In Ref. 1 it is shown that for CSS codes based on a doubly-even classical binary code C with $C_\perp \subset C$, transversal application of the generators of $N(\mathcal{G}_n)$ produces: (i) block-encoded versions of the one-qubit generators; and (ii) a block-encoded CNOT gate. For CSS codes that encode multiple qubits per code block $(k - k' > 1)$, a block-encoded gate applies the gate to all $k - k'$ encoded qubits simultaneously. For such codes, it is not possible to apply an encoded Hadamard or phase gate that acts only on a subset of the encoded qubits, or an encoded CNOT gate that only acts on the encoded i^{th} and j^{th} qubits. However, if the CSS code only encodes a single qubit, this difficulty does not arise and the transversal application of a generator of $N(\mathcal{G}_n)$ yields an encoded fault-tolerant version of this generator. As noted above, combined with the Toffoli gate construction in Sec. 3.3.7, fault-tolerant quantum computation can thus be done on such single-qubit encoding CSS codes. However, to carry out encoded fault-tolerant quantum computing using an arbitrary quantum stabilizer code, more theoretical tools must be developed. Presentation of these new tools begins in the following subsection.

3.3.2. *Quantum Gates Via Measurements*

In this subsection we explain how a quantum gate can be applied to a quantum register using measurements. Transversal application of this technique, in combination with the results of Sec. 3.3.1, will eventually provide a sufficiently powerful set of tools to allow fault-tolerant quantum computation to be done using any quantum stabilizer code.

Let $|\psi_0\rangle$ be an arbitrary state in the n-qubit Hilbert space H_2^n. The set of operators in $\mathcal{G}_n$ that fix $|\psi_0\rangle$ is the stabilizer $\mathcal{S}_0$ of $|\psi_0\rangle$ in $\mathcal{G}_n$. Denoting the generators of $\mathcal{S}_0$ by $\{G_1^0, \ldots, G_m^0\}$, we have

$$G_k^0 |\psi_0\rangle = |\psi_0\rangle \qquad k = 1, \ldots, m. \tag{58}$$

Note that typically $\mathcal{S}_0$ will *not* be the code stabilizer $\mathcal{S}$.

Let $\mathcal{O}$ be an Hermitian operator in $\mathcal{G}_n$ that anticommutes with one of the generators of $\mathcal{S}_0$. We choose the generators so that G_1^0 is the only anticommuting generator. This can always be done since if G_k^0 anticommutes with $\mathcal{O}$, $G_1^0 G_k^0$ will commute with it, and so we simply replace G_k^0 with $G_1^0 G_k^0$. It is a simple matter to show that, since $\mathcal{O}$ is an element of $\mathcal{G}_n$, $\mathcal{O}^2 = I$ and so $\mathcal{O}$ has eigenvalues $\lambda_\pm = \pm 1$ and projection operators $P_\pm = (I \pm \mathcal{O})/2$ that project a state onto the $\lambda_\pm$ eigenspaces, respectively.

We are now ready to introduce the measurement protocol. *Step 1:* Measure $\mathcal{O}$. (A fault-tolerant procedure for measurement is given in Ref. 5.) *Step 2A:* If the measurement outcome is λ_+, take no further action. In this case the final state will be $|\psi_1\rangle = P_+ |\psi_0\rangle \equiv |\psi_+\rangle$. *Step 2B:* If the outcome is λ_-, apply G_1^0 to the post-measurement state $P_- |\psi_0\rangle$. The final state is then

$$\begin{aligned}
|\psi_1\rangle &= G_1^0 P_- |\psi_0\rangle \\
&= G_1^0 \left[\frac{1}{2} (I - \mathcal{O}) \right] |\psi_0\rangle \\
&= \frac{1}{2} (I + \mathcal{O}) G_1^0 |\psi_0\rangle \\
&= |\psi_+\rangle,
\end{aligned} \tag{59}$$

where we have used that G_1^0 anticommutes with $\mathcal{O}$ and fixes $|\psi_0\rangle$. Thus in *both* cases, the measurement protocol maps $|\psi_0\rangle \to |\psi_1\rangle = |\psi_+\rangle$. Following

Gottesman[24] we will say "measure $\mathcal{O}$" when we actually mean "carry out the measurement protocol using $\mathcal{O}$."

Notice that the final state $|\psi_1\rangle = |\psi_+\rangle$ is fixed by the stabilizer $\mathcal{S}_1$ whose generators are $\{\mathcal{O}, G_2^1, \ldots, G_m^1\}$, where

$$G_k^1 = \begin{cases} G_k^0 & \text{if} & [G_k^0, \mathcal{O}] = 0 \\ G_1^0 G_k^0 & \text{if} & \{G_k^0, \mathcal{O}\} = 0 \end{cases} \qquad k \neq 1. \qquad (60)$$

As with $\mathcal{S}_0$, typically $\mathcal{S}_1$ differs from the code stabilizer $\mathcal{S}$. Not surprisingly, since the measurement protocol transforms the state $|\psi_0\rangle \to |\psi_1\rangle$, it also transforms the stabilizer $\mathcal{S}_0 \to \mathcal{S}_1$.

The centralizer $\mathcal{C}(\mathcal{S}_0)$ $(\mathcal{C}(\mathcal{S}_1))$ is the set of operators in $\mathcal{G}_n$ that commute with all elements of $\mathcal{S}_0$ $(\mathcal{S}_1)$. The measurement protocol induces a map from $\mathcal{C}(\mathcal{S}_0)/\mathcal{S}_0 \to \mathcal{C}(\mathcal{S}_1)/\mathcal{S}_1$. To see this, let $N \in \mathcal{C}(\mathcal{S}_0)$ and $|\psi_0\rangle$ the state fixed by $\mathcal{S}_0$. The image N' of N under measurement is determined by requiring that N' acting on the final state $P_+|\psi_0\rangle$ yield the same state as P_+ acting on the state $N|\psi_0\rangle$:

$$N' [P_+|\psi_0\rangle] = P_+ [N|\psi_0\rangle]. \qquad (61)$$

(1) If $[N, \mathcal{O}] = 0$, then $N' = N$ solves eq. (61) since

$$\begin{aligned} N' P_+|\psi_0\rangle &= N P_+|\psi_0\rangle \\ &= P_+ N|\psi_0\rangle. \end{aligned} \qquad (62)$$

Since N commutes with $\mathcal{O}$, $N' \in \mathcal{C}(\mathcal{S}_1)$ and belongs to the coset $N\mathcal{S}_1$. In this case, measurement maps $N \to N'$ and $N\mathcal{S}_0 \to N\mathcal{S}_1$. (2) On the other hand, if $\{N, \mathcal{O}\} = 0$, then $N' = G_1^0 N$ solves eq. (61) since $[G_1^0 N, \mathcal{O}] = 0$ and G_1^0 fixes $|\psi_0\rangle$:

$$\begin{aligned} N' P_+|\psi_0\rangle &= G_1^0 N P_+|\psi_0\rangle \\ &= P_+ G_1^0 N|\psi_0\rangle \\ &= P_+ N|\psi_0\rangle. \end{aligned} \qquad (63)$$

Since $G_1^0 N$ commutes with $\mathcal{O}$, $N' \in \mathcal{C}(\mathcal{S}_1)$ and belongs to the coset $(G_1^0 N)\mathcal{S}_1$. In this case, measurement maps $N \to G_1^0 N$ and $N\mathcal{S}_0 \to (G_1^0 N)\,\mathcal{S}_1$. In the following, the analogs for $\mathcal{S}_0$ of the encoded Pauli operators X_i and Z_i will be denoted $\mathcal{X}_i^0$ and $\mathcal{Z}_i^0$, respectively. By examining how measurement maps $\mathcal{X}_i^0$ and $\mathcal{Z}_i^0$ we will be able to identify what operation is applied to $|\psi_0\rangle$.

In summary, to formally determine the action of measuring $\mathcal{O}$ on the pre-measurement state $|\psi_0\rangle$ (with stabilizer $\mathcal{S}_0$), carry out the following procedure:

(1) Identify a generator $G_1^0 \in \mathcal{S}_0$ that anticommutes with $\mathcal{O}$. Replace any generator G_k^0 ($k \neq 1$) of $\mathcal{S}_0$ that anticommutes with $\mathcal{O}$ with $G_1^0 G_k^0$.
(2) Form the new stabilizer $\mathcal{S}_1$ by replacing G_1^0 in $\mathcal{S}_0$ with $\mathcal{O}$. The generators for $\mathcal{S}_1$ are then $\{\mathcal{O}, G_2^1, \ldots, G_m^1\}$, where the G_k^1 are defined by eq. (60).
(3) Replace any $\mathcal{X}_i^0$ or $\mathcal{Z}_i^0$ that anticommutes with $\mathcal{O}$ with $G_1^0 \mathcal{X}_i^0$ or $G_1^0 \mathcal{Z}_i^0$, respectively. This insures that the new operators belong to $\mathcal{C}(\mathcal{S}_1)$.

The following example due to Gottesman[24] demonstrates how to implement the generators of the Clifford group $N(\mathcal{G}_n)$ using only measurement and a CNOT gate. Since the CNOT gate is assumed available, we only need to show how to implement the Hadamard and phase gates. We now show how to implement the phase gate P, and refer the reader to Refs. 24 and 1 for a demonstration of how to produce the Hadamard gate.

Example 3.2. Consider a two-qubit system in which qubit 1 is in an arbitrary state $|\psi\rangle$ and qubit 2 is in the CB state $|0\rangle$. The stabilizer for the state $|\psi\rangle \otimes |0\rangle$ has generator $G = I^1 \otimes \sigma_z^2$. The operators $\mathcal{X} = \sigma_x^1 \otimes I^2$ and $\mathcal{Z} = \sigma_z^1 \otimes I^2$ clearly commute with G and anticommute with each other. They will serve as the analogs of σ_x and σ_z. Now apply a CNOT gate using qubit 1 as the control. From eq. (56) we have

$$G \rightarrow G^0 = \sigma_z^1 \otimes \sigma_z^2$$
$$\mathcal{X} \rightarrow \mathcal{X}^0 = \sigma_x^1 \otimes \sigma_x^2$$
$$\mathcal{Z} \rightarrow \mathcal{Z}^0 = \sigma_z^1 \otimes I^2. \tag{64}$$

Finally, measure $\mathcal{O} = I^1 \otimes \sigma_y^2$. It is clear from eq. (64) that G^0 and $\mathcal{X}^0$ anticommute with $\mathcal{O}$. Thus measuring $\mathcal{O}$ induces the map

$$G^0 \rightarrow G^1 = \mathcal{O} = I^1 \otimes \sigma_y^2$$
$$\mathcal{X}^0 \rightarrow \mathcal{X}^1 = G^0 \mathcal{X}^0 = -\sigma_y^1 \otimes \sigma_y^2$$
$$\mathcal{Z}^0 \rightarrow \mathcal{Z}^1 = \sigma_z^1 \otimes I^2. \tag{65}$$

Since the measurement protocol leaves qubit 2 in the +1 eigenstate of σ_y^2, the parts of G^1, $\mathcal{X}^1$, and $\mathcal{Z}^1$ associated with qubit 2 fix this state and so it is conventional to discard the measured qubit. The parts of $\mathcal{X}^0$, $\mathcal{X}^1$, $\mathcal{Z}^0$, and $\mathcal{Z}^1$ associated with qubit 1 then determine what operation has been carried out. From $\mathcal{X}^0 \rightarrow \mathcal{X}^1$ we see that $\sigma_x^1 \rightarrow -\sigma_y^1$, and from $\mathcal{Z}^0 \rightarrow \mathcal{Z}^1$ that $\sigma_z^1 \rightarrow \sigma_z^1$. The above procedure has thus implemented $P^\dagger$ (see eq. (55)). Note that if the measurement protocol were modified so that $|\psi_0\rangle \rightarrow P_-|\psi_0\rangle$, the above procedure would have produced P instead of $P^\dagger$.

As noted above, it is possible to construct the Hadamard gate H using measurement and a CNOT gate. Thus all generators in the Clifford group can be implemented having access only to measurement and a CNOT gate. It is important to note that these gates are unencoded. Before going on to discuss how to produce encoded versions of the Clifford group generators, we take a brief detour to introduce a four-qubit unitary operation U_4 in the following subsection which will prove to be extremely useful.

3.3.3. *Four-Qubit Operation U_4*

This subsection introduces a four-qubit unitary operation that, when applied transversally, produces an encoded version of itself for *all* quantum stabilizer codes. With this operation, a fault-tolerant block-encoded CNOT gate can be made, which together with measurement, allows all gates in the Clifford group to be constructed (see Sec. 3.3.2) for stabilizer codes encoding a single-qubit.

We begin by specifying the action of the four-qubit operation U_4. It produces the following transformation on the generators of $\mathcal{G}_4$:

$$
\begin{aligned}
\sigma_x^1 \otimes I^2 \otimes I^3 \otimes I^4 &\longrightarrow \sigma_x^1 \otimes \sigma_x^2 \otimes \sigma_x^3 \otimes I^4 \\
I^1 \otimes \sigma_x^2 \otimes I^3 \otimes I^4 &\longrightarrow I^1 \otimes \sigma_x^2 \otimes \sigma_x^3 \otimes \sigma_x^4 \\
I^1 \otimes I^2 \otimes \sigma_x^3 \otimes I^4 &\longrightarrow \sigma_x^1 \otimes I^2 \otimes \sigma_x^3 \otimes \sigma_x^4 \\
I^1 \otimes I^2 \otimes I^3 \otimes \sigma_x^4 &\longrightarrow \sigma_x^1 \otimes \sigma_x^2 \otimes I^3 \otimes \sigma_x^4 \\[6pt]
\sigma_z^1 \otimes I^2 \otimes I^3 \otimes I^4 &\longrightarrow \sigma_z^1 \otimes \sigma_z^2 \otimes \sigma_z^3 \otimes I^4 \\
I^1 \otimes \sigma_z^2 \otimes I^3 \otimes I^4 &\longrightarrow I^1 \otimes \sigma_z^2 \otimes \sigma_z^3 \otimes \sigma_z^4 \\
I^1 \otimes I^2 \otimes \sigma_z^3 \otimes I^4 &\longrightarrow \sigma_z^1 \otimes I^2 \otimes \sigma_z^3 \otimes \sigma_z^4 \\
I^1 \otimes I^2 \otimes I^3 \otimes \sigma_z^4 &\longrightarrow \sigma_z^1 \otimes \sigma_z^2 \otimes I^3 \otimes \sigma_z^4.
\end{aligned}
\tag{66}
$$

Let $\mathcal{C}_q$ be a quantum stabilizer code that encodes k qubits in n qubits. Let

$$
s = i^{\lambda_s} \sigma_x(s_x) \sigma_z(s_z)
$$

be an arbitrary element of the stabilizer $\mathcal{S}$. Here λ_s is chosen to insure s is Hermitian, and s_x and s_z are bit strings of length n. Suppose we apply the four-qubit operation U_4 transversally to four code blocks, with each of the Pauli operators in s transformed according to eq. (66). This operation produces the mapping:

$$s \otimes I^2 \otimes I^3 \otimes I^4 \longrightarrow s \otimes s \otimes s \otimes I^4$$
$$I^1 \otimes s \otimes I^3 \otimes I^4 \longrightarrow I^1 \otimes s \otimes s \otimes s$$
$$I^1 \otimes I^2 \otimes s \otimes I^4 \longrightarrow s \otimes I^2 \otimes s \otimes s$$
$$I^1 \otimes I^2 \otimes I^3 \otimes s \longrightarrow s \otimes s \otimes I^3 \otimes s. \tag{67}$$

The reader can show that the generator images are themselves generators of the stabilizer $\mathcal{S} \otimes \mathcal{S} \otimes \mathcal{S} \otimes \mathcal{S}$ by finding combinations of their products that yield the original set of generators. This operation thus fixes the four-block stabilizer and so is an encoded operation. Since it is transversal it is also fault-tolerant. To determine which encoded operation it is, we examine how the encoded Pauli operators $X_j = i^{\lambda_j} \sigma_x(c_j) \sigma_z(d_j)$ and $Z_j = i^{\lambda'_j} \sigma_x(e_j) \sigma_z(f_j)$ $(j = 1, \ldots, k)$ are transformed. Transversal application of U_4 maps:

$$X_j \otimes I^2 \otimes I^3 \otimes I^4 \longrightarrow X_j \otimes X_j \otimes X_j \otimes I^4$$
$$I^1 \otimes X_j \otimes I^3 \otimes I^4 \longrightarrow I^1 \otimes X_j \otimes X_j \otimes X_j$$
$$I^1 \otimes I^2 \otimes X_j \otimes I^4 \longrightarrow X_j \otimes I^2 \otimes X_j \otimes X_j$$
$$I^1 \otimes I^2 \otimes I^3 \otimes X_j \longrightarrow X_j \otimes X_j \otimes I^3 \otimes X_j$$

$$Z_j \otimes I^2 \otimes I^3 \otimes I^4 \longrightarrow Z_j \otimes Z_j \otimes Z_j \otimes I^4$$
$$I^1 \otimes Z_j \otimes I^3 \otimes I^4 \longrightarrow I^1 \otimes Z_j \otimes Z_j \otimes Z_j$$
$$I^1 \otimes I^2 \otimes Z_j \otimes I^4 \longrightarrow Z_j \otimes I^2 \otimes Z_j \otimes Z_j$$
$$I^1 \otimes I^2 \otimes I^3 \otimes Z_j \longrightarrow Z_j \otimes Z_j \otimes I^3 \otimes Z_j, \tag{68}$$

with $j = 1, \ldots, k$. Comparing eqs. (66) and (68) we see that transversal application of the four-qubit operation U_4 produces an encoded version of itself. Since the code $\mathcal{C}_q$ was chosen arbitrarily, this result is true for *all* quantum stabilizer codes. We see from eq. (68) that for multi-qubit encodings $(k > 1)$, the encoded U_4 operation *simultaneously* maps the encoded Pauli operators (X_j, Z_j) for *all* j. It cannot be used to map a subset of these operators (e. g. only X_1 and Z_1). Clearly, for single-qubit encodings $(k = 1)$, this issue does not arise. We now show how U_4 and measurement can be used to produce a CNOT gate.

3.3.4. *CNOT Gate Using U_4 and Measurement*

Now suppose we have two qubits in an arbitrary two-qubit state $|\psi_{12}\rangle$ and we prepare two ancilla qubits in the CB state $|00\rangle$. The stabilizer $\mathcal{S}_0$ for the state $|\psi_{12}\rangle \otimes |00\rangle$ has generators:

$$G_1^0 = I^1 \otimes I^2 \otimes \sigma_z^3 \otimes I^4 \quad ; \quad G_2^0 = I^1 \otimes I^2 \otimes I^3 \otimes \sigma_z^4, \qquad (69)$$

and the Pauli operator analogs are

$$\mathcal{X}_1^0 = \sigma_x^1 \otimes I^2 \otimes I^3 \otimes I^4 \quad ; \quad \mathcal{X}_2^0 = I^1 \otimes \sigma_x^2 \otimes I^3 \otimes I^4$$
$$\mathcal{Z}_1^0 = \sigma_z^1 \otimes I^2 \otimes I^3 \otimes I^4 \quad ; \quad \mathcal{Z}_2^0 = I^1 \otimes \sigma_z^2 \otimes I^3 \otimes I^4. \qquad (70)$$

Applying U_4 to these qubits produces the mapping:

$$G_1^0 \to G_1 = \sigma_z^1 \otimes I^2 \otimes \sigma_z^3 \otimes \sigma_z^4 \quad ; \quad G_2^0 \to G_2 = \sigma_z^1 \otimes \sigma_z^2 \otimes I^3 \otimes \sigma_z^4$$
$$\mathcal{X}_1^0 \to \mathcal{X}_1 = \sigma_x^1 \otimes \sigma_x^2 \otimes \sigma_x^3 \otimes I^4 \quad ; \quad \mathcal{X}_2^0 \to \mathcal{X}_2 = I^1 \otimes \sigma_x^2 \otimes \sigma_x^3 \otimes \sigma_x^4$$
$$\mathcal{Z}_1^0 \to \mathcal{Z}_1 = \sigma_z^1 \otimes \sigma_z^2 \otimes \sigma_z^3 \otimes I^4 \quad ; \quad \mathcal{Z}_2^0 \to \mathcal{Z}_2 = I^1 \otimes \sigma_z^2 \otimes \sigma_z^3 \otimes \sigma_z^4.$$

Finally, the operators $\mathcal{O}_1 = I^1 \otimes I^2 \otimes \sigma_x^3 \otimes I^4$ and $\mathcal{O}_2 = I^1 \otimes I^2 \otimes I^3 \otimes \sigma_x^4$ are measured. They anticommute, respectively, with $\{G_1, \mathcal{Z}_1, \mathcal{Z}_2\}$ and $\{G_1, G_2, \mathcal{Z}_2\}$. Measurement thus produces the mapping:

$$G_1, G_2 \to G_1^1, G_2^1$$
$$\mathcal{X}_1, \mathcal{X}_2 \to \mathcal{X}_1^1, \mathcal{X}_2^1$$
$$\mathcal{Z}_1, \mathcal{Z}_2 \to \mathcal{Z}_1^1, \mathcal{Z}_2^1,$$

where $\mathcal{Z}_1^1 = G_1 G_2 \mathcal{Z}_1$; $\mathcal{Z}_2^1 = G_1 \mathcal{Z}_2$; and

$$G_1^1 = \mathcal{O}_1 \quad ; \quad G_2^1 = \mathcal{O}_2$$
$$\mathcal{X}_1^1 = \sigma_x^1 \otimes \sigma_x^2 \otimes \sigma_x^3 \otimes I^4 \quad ; \quad \mathcal{X}_2^1 = I^1 \otimes \sigma_x^2 \otimes \sigma_x^3 \otimes \sigma_x^4$$
$$\mathcal{Z}_1^1 = \sigma_z^1 \otimes I^2 \otimes I^3 \otimes I^4 \quad ; \quad \mathcal{Z}_2^1 = \sigma_z^1 \otimes \sigma_z^2 \otimes I^3 \otimes I^4.$$

Discarding the measured ancilla qubits gives

$$\mathcal{X}_1^1 = \sigma_x^1 \otimes \sigma_x^2 \quad ; \quad \mathcal{X}_2^1 = I^1 \otimes \sigma_x^2$$
$$\mathcal{Z}_1^1 = \sigma_z^1 \otimes I^2 \quad ; \quad \mathcal{Z}_2^1 = \sigma_z^1 \otimes \sigma_z^2. \qquad (71)$$

Discarding the ancilla qubits from eq. (70) we see that this procedure has mapped $\mathcal{X}_1^0, \mathcal{X}_2^0 \to \mathcal{X}_1^1, \mathcal{X}_2^1$ and $\mathcal{Z}_1^0, \mathcal{Z}_2^0 \to \mathcal{Z}_1^1, \mathcal{Z}_2^1$, which (restricted to the first two qubits) maps:

$$\sigma_x^1 \otimes I^2 \longrightarrow \sigma_x^1 \otimes \sigma_x^2$$
$$I^1 \otimes \sigma_x^2 \longrightarrow I^1 \otimes \sigma_x^2$$
$$\sigma_z^1 \otimes I^2 \longrightarrow \sigma_z^1 \otimes I^2$$
$$I^1 \otimes \sigma_z^2 \longrightarrow \sigma_z^1 \otimes \sigma_z^2. \qquad (72)$$

We recognize eq. (72) as the action of a CNOT gate on the generators of $\mathcal{G}_2$. Thus U_4 and an appropriate measurement allow us to apply a CNOT gate to a pair of qubits. We now extend this procedure to produce a block-encoded CNOT gate.

3.3.5. *Block Encoded CNOT Using U_4 and Measurement*

If we apply the procedure of Sec. 3.3.4 transversally to four code blocks and discard the measured qubits in the ancilla code blocks, we produce a fault-tolerant encoded operation on the first two blocks that maps

$$
\begin{aligned}
X_j \otimes I^2 &\longrightarrow X_j \otimes X_j \\
I^1 \otimes X_j &\longrightarrow I^1 \otimes X_j \\
Z_j \otimes I^2 &\longrightarrow Z_j \otimes I^2 \\
I^1 \otimes Z_j &\longrightarrow Z_j \otimes Z_j,
\end{aligned}
\tag{73}
$$

for $j = 1, \ldots, k$. We recognize this as a block-encoded CNOT gate. As with eq. (68), for stabilizer codes encoding more than one qubit, this procedure causes an encoded CNOT gate to be applied simultaneously to the jth encoded qubits in the two code blocks, for all j values. It cannot be used to selectively apply an encoded CNOT gate to some of the j, and not to others. This procedure is also unable to apply an encoded CNOT gate to two encoded qubits in the *same* code block. This difficulty does not occur if the stabilizer code only encodes a single qubit. For such codes, the procedure yields a fault-tolerant encoded CNOT gate, which together with measurement allows all gates in the Clifford group to be applied fault-tolerantly on encoded qubits. The following subsection extends this procedure so that the same can be done with stabilizer codes that encode multiple qubits.

3.3.6. *Clifford Group Gates—Multi-Qubit Quantum Stabilizer Codes*

Section 3.3.5 showed how the four-qubit operation U_4, used in conjunction with measurement, could be used to construct a block-encoded CNOT gate. We saw that for stabilizer codes that encode multiple qubits, this block CNOT gate was unable to apply an encoded CNOT gate: (i) between only the ith encoded qubits; or (ii) between only the ith and j^{th} encoded qubits in the same or different blocks. This subsection presents techniques that remove these limitations so that it will be possible to fault-tolerantly perform all gates in the Clifford group on encoded qubits for any quantum stabilizer code.

To begin, suppose we have a data qubit and an ancilla qubit. The data qubit is initially in an arbitrary state $|\psi\rangle$ and we prepare the ancilla qubit in the $+1$ eigenstate of σ_x^a. The stabilizer for this state has the generator $G^0 = I^d \otimes \sigma_x^a$, where the superscripts identify the data and ancilla qubit operators. The Pauli operator analogs are $\mathcal{X}^0 = \sigma_x^d \otimes I^a$ and $\mathcal{Z}^0 = \sigma_z^d \otimes I^a$. A CNOT gate is now applied to this state using the ancilla qubit as the control. This produces the new generator and Pauli operator analogs: $\overline{G} = \sigma_x^d \otimes \sigma_x^a$, $\overline{\mathcal{X}} = \sigma_x^d \otimes I^a$, and $\overline{\mathcal{Z}} = \sigma_z^d \otimes \sigma_z^a$. Finally, we measure $\mathcal{O} = \sigma_z^d \otimes I^a$ so that the final generator and Pauli operator analogs are $G^1 = \mathcal{O}$, and $\mathcal{X}^1 = I^d \otimes \sigma_x^a$ and $\mathcal{Z}^1 = \sigma_z^d \otimes \sigma_z^a$, respectively. Discarding the measured qubit gives $\mathcal{X}^1 = \sigma_x^a$ and $\mathcal{Z}^1 = \sigma_z^a$ so that $\sigma_x^d \to \sigma_x^a$ and $\sigma_z^d \to \sigma_z^a$. Thus the data has been transferred from the data qubit to the ancilla qubit. Now suppose that we had initially prepared the ancilla qubit in the $+1$ eigenstate of σ_z^a (viz. CB state $|0\rangle$) and then applied the CNOT gate using the ancilla as the control. Since the control qubit is in the $|0\rangle$ state, no data is transferred because the CNOT gate applies no action to the target qubit for this control state.

The two operations introduced in the preceding paragraph provide the means to move the jth encoded qubit from a data code block to an empty ancilla code block. We prepare the ancilla block so that all encoded qubits $i \neq j$ are in the $+1$ eigenstate $|\overline{0}\rangle$ of Z_i^a and the jth encoded qubit is in the $+1$ eigenstate of X_j^a. We then apply the block encoded CNOT gate from Sec. 3.3.5, using the ancilla block as the control and then measure Z_j^d. This produces the encoded version of the two operations discussed in the preceding paragraph so that the jth encoded qubit is transferred to the jth encoded ancilla qubit, while all other encoded ancilla qubits remain in the state $|\overline{0}\rangle$. After the measurement, the jth encoded data qubit is left in the $+1$ eigenstate $|\overline{0}\rangle$ of Z_j^d, while no action is applied to the other encoded data qubits. This procedure thus provides the means to switch a specific encoded qubit from one code block to another. The procedure to transfer the jth encoded qubit back to the data block is to apply a block-encoded CNOT gate using the ancilla block as the control and then measure X_j^a. It can be shown[1] that this produces the mapping $X_j^a \to X_j^d$ and $Z_j^a \to Z_j^d$ so that the state of the encoded ancilla qubit is transferred to the encoded data qubit.

We can now overcome the first limitation mentioned above—applying an encoded CNOT gate solely to the ith encoded qubits in two data blocks. To do this we first switch the ith encoded qubits to two ancilla code blocks using the procedure presented in the previous paragraph. Upon completion of this procedure, the ith encoded qubits have been transferred to the two ancilla blocks and all other encoded ancilla qubits are in the state $|\overline{0}\rangle$. We can now apply the block-encoded CNOT gate from Sec. 3.3.5 to the ancilla blocks. This will apply an encoded CNOT gate to the ith encoded qubits while leaving unaltered the state of the other encoded ancilla qubits. At this point we have applied the desired operation to the ith encoded qubits and so can switch them back to their initial data blocks by the procedure described at the end of the previous paragraph.

The remaining limitation is how to apply an encoded CNOT gate to the ith and j^{th} encoded qubits in either the same or different data blocks. A partial solution is to transfer these encoded data qubits to two ancilla blocks in which all other encoded ancilla qubits are in the state $|\bar{0}\rangle$. If a fault-tolerant encoded SWAP gate is available to swap a given encoded qubit with the first encoded qubit in its block, we could then swap each of the two encoded data qubits with the first encoded qubit in its respective block. Applying the block encoded CNOT gate to the ancilla blocks would apply an encoded CNOT gate to the two encoded data qubits while leaving all other encode ancilla qubits alone. Having applied the desired CNOT gate to the encoded data qubits, we could swap them back to their starting positions in the ancilla blocks, and then switch them back to their initial data blocks. Clearly, for this approach to work, a procedure must be found to fault-tolerantly apply the desired encoded SWAP gate. Ref. 1 explains how such a gate can be carried out using quantum teleportation. Combining all the procedures presented in this subsection we can now fault-tolerantly apply encoded versions of all operations in the Clifford group $N(\mathcal{G}_n)$ for any quantum stabilizer code. Together with the encoded Toffoli gate introduced in the following subsection, we will finally be able to do fault-tolerant quantum computing on data encoded using any quantum stabilizer code.

3.3.7. *Toffoli Gate*

Using the techniques of Secs. 3.3.1-3.3.6, we can now apply all gates in the Clifford group $N(\mathcal{G}_n)$ fault-tolerantly for any quantum stabilizer code. Unfortunately, this set of gates is not sufficiently powerful to perform universal quantum computation. The Gottesman-Knill theorem[1] gives a careful statement and proof of this. Universal quantum computation becomes possible, however, if the gates in the Clifford group are supplemented with the Toffoli gate.[16,20] This subsection completes the theoretical framework that allows fault-tolerant quantum computing to be done using any quantum stabilizer code by showing how to make a fault-tolerant encoded Toffoli gate.[16,24] Note that other gates can be used instead of the Toffoli gate. A well-known alternative is the $\pi/8$ gate. We refer the reader to Refs. 25 and 26 for a discussion of its fault-tolerant encoded implementation.

The Toffoli gate U_T is a controlled-controlled NOT gate that uses qubits 1 and 2 as the controls and qubit 3 as the target. Its action on the three-qubit CB state $|i_1 i_2 i_3\rangle$ is to apply a NOT gate to qubit 3 only when $i_1 = i_2 = 1$. Otherwise, no action is applied to the qubits. Formally, the action of U_T on the three-qubit CB states is

$$U_T |i_1 i_2 i_3\rangle = |i_1 i_2 (i_3 \oplus i_1 i_2)\rangle, \tag{74}$$

where $\oplus$ denotes binary addition and $i_1, i_2, i_3 = 0, 1$. In terms of the projection operators $P_\pm(\sigma_z) = (I \pm \sigma_z)/2$, U_T can be written as

$$
\begin{aligned}
U_T &= \frac{1}{2} \left\{ P_+(\sigma_z^1) + P_+(\sigma_z^2) + P_-(\sigma_z^1 \sigma_z^2) \right\} + P_-(\sigma_z^1) P_-(\sigma_z^2) \sigma_x^3 \\
&= \frac{1}{4} \left[3I + \sigma_z^1 + \sigma_z^2 - \sigma_z^1 \sigma_z^2 + \left(I - \sigma_z^1 \right) \left(I - \sigma_z^2 \right) \sigma_x^3 \right].
\end{aligned}
\tag{75}
$$

In an effort to make equations more readable, we will suppress the direct product symbol $\otimes$ throughout this subsection.

U_T acts on the three-qubit Pauli group $\mathcal{G}_3$ by conjugation. Its action thus preserves the multiplication table of $\mathcal{G}_3$ and we can focus on how U_T maps the generators of $\mathcal{G}_3$. We shall see momentarily that the image of $\mathcal{G}_3$ under U_T is contained in $N(\mathcal{G}_3)$. Consequently, the Toffoli gate U_T does not fix $\mathcal{G}_3$ and so cannot belong to $N(\mathcal{G}_3)$. It cannot belong to the Clifford group $N(\mathcal{G}_n)$ either since, if it did, it would fix $\mathcal{G}_n$ and so would have to fix the subgroup $\mathcal{G}_3 \subset \mathcal{G}_n$. Since it does not, the Toffoli gate has to lie outside the Clifford group. It is possible to show that U_T carries out the following map of the generators of $\mathcal{G}_3$[1]

$$
\begin{aligned}
\sigma_x^1 &\to U_{CNOT}^{23} \sigma_x^1 \equiv G_1 \\
\sigma_x^2 &\to U_{CNOT}^{13} \sigma_x^2 \equiv G_2 \\
\sigma_x^3 &\to \sigma_x^3 \\
\sigma_z^1 &\to \sigma_z^1 \\
\sigma_z^2 &\to \sigma_z^2 \\
\sigma_z^3 &\to U_{CP}^{12} \sigma_z^3 \equiv G_3.
\end{aligned}
\tag{76}
$$

Here U_{CNOT}^{ij} is a CNOT gate from qubit i to qubit j and U_{CP}^{ij} is the controlled-phase gate acting on qubits i and j introduced in Sec. 3.3.1. We see from eq. (76) that the generator images all lie in $N(\mathcal{G}_3)$ since they can be constructed from gates in the Clifford group. Since U_T preserves the multiplication table of $\mathcal{G}_3$, all elements of $\mathcal{G}_3$ are mapped into $N(\mathcal{G}_3)$ as claimed earlier.

Suppose that we can prepare the three-qubit state

$$|A\rangle = \frac{1}{2} \left[|000\rangle + |010\rangle + |100\rangle + |111\rangle \right]. \tag{77}$$

The stabilizer for this state has generators G_1, G_2, and G_3 which were introduced in eq. (76). Now introduce another three-qubit block (with qubits labeled 4, 5, 6) prepared in an arbitrary three-qubit state $|\psi\rangle$. The stabilizer for the composite state $|A\rangle|\psi\rangle$ has generators $G_1^0 = G_1 I^4 I^5 I^6$, $G_2^0 = G_2 I^4 I^5 I^6$, and $G_3^0 = G_3 I^4 I^5 I^6$, and the Pauli operator analogs are $\mathcal{X}_1^0 = \sigma_x^4$, $\mathcal{X}_2^0 = \sigma_x^5$, $\mathcal{X}_3^0 = \sigma_x^6$, and $\mathcal{Z}_1^0 = \sigma_x^4$, $\mathcal{Z}_2^0 = \sigma_z^5$, $\mathcal{Z}_3^0 = \sigma_z^6$. Next apply CNOT gates from qubits $1 \to 4$, $2 \to 5$, and $6 \to 3$, and finally measure $\mathcal{O}_1 = \sigma_z^4$, $\mathcal{O}_2 = \sigma_z^5$, and $\mathcal{O}_3 = \sigma_x^6$ and discard the measured qubits. The result of this procedure is the following mapping:[1]

$$\sigma_x^4 \to G_1$$
$$\sigma_x^5 \to G_2$$
$$\sigma_x^6 \to \sigma_x^3$$
$$\sigma_z^4 \to \sigma_z^1$$
$$\sigma_z^5 \to \sigma_z^2$$
$$\sigma_z^6 \to G_3. \tag{78}$$

We recognize eq. (78) as the action of the Toffoli gate (see eq. (76)) in combination with the transfer of all data from qubits 4-6 to qubits 1-3. Thus if we can produce an encoded $|A\rangle$ state, we can apply the above procedure using encoded CNOT gates and measurements (which we know how to do) to implement an encoded Toffoli gate. To simplify the discussion of making an encoded $|A\rangle$ state, we focus on a quantum stabilizer code that encodes a single qubit. To treat a multi-qubit code we would start with three ancilla blocks with all encoded qubits in the state $|\overline{0}\rangle$ and use the X_1 and Z_1 operators for each block to implement the procedure described below to make the encoded $|A\rangle$ state. At the end of this procedure, the first encoded qubits in these blocks would be in the encoded $|A\rangle$ state and all other encoded qubits would remain in the state $|\overline{0}\rangle$. The three encoded data blocks that are about to be acted on by the Toffoli gate are switched to three ancilla blocks in which all encoded qubits are initially in the state $|\overline{0}\rangle$. The three encoded data qubits are then swapped with the first encoded qubits in their respective blocks. The just-described Toffoli gate procedure is then applied to the six ancilla blocks using the block-encoded CNOT gates and measurements. Afterwards the encoded data qubits are transferred back to their original code blocks. There is thus no loss of generality in focusing on single-qubit encodings.

Let $|\overline{A}\rangle$ represent the encoded version of eq. (77) and introduce $|\overline{B}\rangle = X_3|\overline{A}\rangle$, where X_3 is the encoded σ_x^3 operator. Let the stabilizer code $\mathcal{C}_q$ that encodes each of the three code blocks be an $[n, 1, d]$ code. We will at times below refer to the collection of three code blocks as simply the code block to unclutter the language. It should be clear from the context when this is being done. Suppressing normalization factors in the remainder of this subsection, the states $|\overline{A}\rangle$ and $|\overline{B}\rangle$ are

$$|\overline{A}\rangle = |\overline{000}\rangle + |\overline{010}\rangle + |\overline{100}\rangle + |\overline{111}\rangle$$
$$|\overline{B}\rangle = |\overline{001}\rangle + |\overline{011}\rangle + |\overline{101}\rangle + |\overline{110}\rangle. \tag{79}$$

Notice that

$$|\overline{A}\rangle + |\overline{B}\rangle = \sum_{i_1,i_2,i_3=0}^{1} |\overline{i_1 i_2 i_3}\rangle$$
$$= \left[\, |\overline{0}\rangle + |\overline{1}\rangle \,\right] \left[\, |\overline{0}\rangle + |\overline{1}\rangle \,\right] \left[\, |\overline{0}\rangle + |\overline{1}\rangle \,\right]. \tag{80}$$

The state $|\overline{A}\rangle + |\overline{B}\rangle$ can thus be prepared by measuring X_i on each code block $i = 1, 2, 3$. Next introduce an ancilla block containing n qubits that has been prepared in the cat-state $|\psi_{cat}\rangle = |0\cdots 0\rangle + |1\cdots 1\rangle$. Let $\overline{G}_3$ be the encoded version of G_3 (see eq. (76)), and suppose we can apply a controlled-$\overline{G}_3$ gate to the combined code and ancilla blocks (viz. $\overline{G}_3$ is (not) applied to the code block when the ancilla block is in the state $|1\cdots 1\rangle$ ($|0\cdots 0\rangle$)). With this operation the state $|\overline{A}\rangle$ can be constructed as follows. Start with the ancilla block prepared in the state $|\psi_{cat}\rangle$ and the code block in the state $|\overline{A}\rangle + |\overline{B}\rangle$. Apply the controlled-$\overline{G}_3$ operation to the composite state. The result is

$$[|0\cdots 0\rangle + |1\cdots 1\rangle]\,[|\overline{A}\rangle + |\overline{B}\rangle] \longrightarrow$$
$$[|0\cdots 0\rangle + |1\cdots 1\rangle]\,|\overline{A}\rangle + [|0\cdots 0\rangle - |1\cdots 1\rangle]\,|\overline{B}\rangle. \tag{81}$$

Finally, measure $\mathcal{O} = \sigma_x^1 \cdots \sigma_x^n$ on the ancilla block. Since

$$\mathcal{O}\,[|0\cdots 0\rangle \pm |1\cdots 1\rangle] = \pm\,[|0\cdots 0\rangle \pm |1\cdots 1\rangle],$$

if the measurement outcome is $+1$, we know from eq. (81) that the code block will be in the state $|\overline{A}\rangle$. If the outcome is -1, then the code block is in the state $|\overline{B}\rangle$. In this case, applying the operator X_3 to the code block puts it in the state $X_3|\overline{B}\rangle = |\overline{A}\rangle$. In either case, at the end of all operations, the code block will be in the state $|\overline{A}\rangle$ as desired.

As we know from Sec. 3.2.2, cat-state preparation is not fault-tolerant
and so a candidate cat-state must be carefully verified before it is used
in the production of the state $|\overline{A}\rangle$. The quantum circuit in Fig. 3 (minus
the final four Hadamard gates) insures that there are no bit-flip errors in
the ancilla block state to $\mathcal{O}(p)$, but it does not detect if a phase error occurs. A phase error in the cat-state can be shown[1] to cause an error in
the construction of $|\overline{A}\rangle$. Specifically, the code block will be in the state $|\overline{B}\rangle$
($|\overline{A}\rangle$) when the measurement outcome is $+1$ (-1) instead of $|\overline{A}\rangle$ ($|\overline{B}\rangle$). A
procedure to handle phase errors to $\mathcal{O}(p)$ is described in Ref. 1. We also
show there how the controlled-$\overline{G}_3$ gate can be constructed using transversal operations, measurements, and a universal set of unencoded quantum
gates. With these ingredients, the above discussion shows how we can apply a fault-tolerant encoded Toffoli gate using any quantum stabilizer code.
Putting together all results from Sec. 3, we now have the means to carry out
fault-tolerant quantum computation on data encoded using any quantum
stabilizer code. Having established the theoretical framework for quantum
error correction and fault-tolerant quantum computing, we move on to a
proof of the accuracy threshold theorem in Sec. 4.

4. Accuracy Threshold Theorem

The stage is finally set for a proof of the accuracy threshold theorem.[27-34]
As pointed out in Sec. 1, this theorem spells out the conditions under
which quantum computing can be done reliably using imperfect quantum
gates and in the presence of noise. The proof described in this Sec. follows
Ref. 31. Section 4.1 addresses a number of preliminary topics. It explains
why a concatenated quantum error correcting code (QECC) should be used
to protect the computational data when fault-tolerance is an issue; it states
the principal assumptions underlying the threshold calculations in Sec. 4.3;
and finally, it closes with a statement of the accuracy threshold theorem.
The proof of the theorem is spread out over Secs. 4.2 and 4.3. For the error
model introduced in Sec. 4.1, the accuracy threshold is calculated for gates
in the Clifford group $N(\mathcal{G}_n)$, storage registers, and the Toffoli gate. The
calculations make explicit use of the recursive structure of a concatenated
QECC and the fault-tolerant procedures introduced in Sec. 3. The theorem
is proved by showing that all three cases yield non-zero threshold values.

4.1. *Preliminaries*

We have learned how to protect data using a QECC and how to control the proliferation of errors through the use of fault-tolerant procedures for encoded quantum gates, error correction, and measurement. The question that faces us now is whether this substantial theoretical investment provides us with the ability to do an arbitrary quantum computation with arbitrarily small error probability. Ultimately, we will find the answer is yes, though only under appropriate conditions. First, it is clear that active harm will be done to the data if an error correction step introduces more errors than it removes. Thus one anticipates that if error correction is to be useful, the error probabilities for quantum gates and storage registers used for quantum computation and error correction cannot be too large. Second, and perhaps a little less obvious, is that not all QECCs are suitable for fault-tolerant quantum computing. The introduction of fault-tolerant procedures increases both the circuit complexity and time needed to do syndrome extraction. Ironically, the desire to control error proliferation through fault-tolerant procedures provides more opportunities for errors to be introduced into the data! Again, if error probabilities are too large, more errors will appear during error correction than are removed. Concatenated QECCs provide a way around this second difficulty. Their recursive structure allows them to correct a large number of errors, while at the same time making it possible to do syndrome extraction in a sufficiently efficient manner.

4.1.1. *Concatenated Quantum Error Correcting Codes*

Concatenated QECCs were introduced in Sec. 2.4. There, we discussed codes with only two layers of concatenation. In this section we will be interested in codes with l layers of concatenation.

In situations where fault-tolerance is an issue, it is highly convenient when constructing a concatenated QECC to base the construction on a CSS code that uses a doubly-even classical code with $C_\perp \subset C$ (Sec. 3.3.1). The reason such a code is so convenient is that encoded gates in the Clifford group can be applied fault-tolerantly by applying the unencoded gates transversally to code blocks. Consequently, for the remainder of Sec. 4, we restrict ourselves to a concatenated QECC that is based on the [7,1,3] CSS code which is a based on the doubly-even [7,4,3] binary Hamming code.

Concatenated QECCs can be given two complimentary descriptions: (i) top-down; and (ii) bottom-up.

(i) In the top-down description we begin with an unencoded qubit and encode it into a block of seven qubits using the [7,1,3] CSS code. Next each qubit in this block is itself encoded into a block of seven qubits. Repeating this process l times produces our concatenated QECC in which one data qubit is recursively encoded into the state of 7^l physical qubits. It is possible to show[1] that our concatenated code is a $[7^l, 1, d]$ code with $d \geq 3^l$. Recall that a distance d QECC can correct t errors, where $t = [(d-1)/2]$ and $[x]$ is the integer part of x. Since $d \geq 3^l$, we see that the number of errors that a concatenated QECC can correct grows *exponentially* with the number of concatenation layers l.

(ii) In the remainder of Sec. 4 we will use the bottom-up description of a concatenated QECC. In the bottom-up description we begin with 7^l physical qubits. This is the zeroth-level of the (concatenated) code. The first-level of the code is obtained by collecting the physical qubits into non-overlapping blocks of seven qubits. Each block encodes a single level-1 qubit using the [7,1,3] CSS code. There are thus a total of 7^{l-1} such level-1 qubits. The second-level of the code is obtained by collecting the level-1 qubits into (non-overlapping) blocks of seven qubits so that there are 7^{l-2} such level-2 qubits. The blocking/coarse-graining of level-$(j-1)$ qubits to form level-j qubits is repeated l times so that at level-l there is only one level-l qubit, which is the unencoded data qubit we are trying to protect. From the construction procedure it follows that a level-j qubit is made up of 7^j physical qubits. To measure the error syndrome for this code we begin at level-1, measuring the syndrome for each level-1 code block. To simplify the analysis of Secs. 4.2 and 4.3 it is assumed that the syndrome for all these code blocks can be measured in parallel. Then we do not need to worry about storage errors accumulating on code blocks while they wait around for other code blocks to have their syndrome measured. As long as l is not too large, this assumption is reasonable, and so the time to measure the level-1 error syndrome is simply the time it takes to measure and verify the syndrome of a level-1 (code) block. Once the syndrome for all level-1 blocks has been determined, the appropriate recovery operation is applied in parallel to each block. Next, the error syndrome is measured in parallel for all level-2 blocks. As with the level-1 blocks, the time needed to do this is the time it takes to measure and verify the error syndrome for a level-2 block. Once the syndrome has been determined for all level-2 blocks, error recovery is done in parallel on each of these blocks. Repeating this process for all l code-levels implements one error correction cycle on our concatenated QECC. The time required to carry out one error correction cycle is then the sum of the times needed to do error correction on a code block at each code-level. We can now state why concatenated QECCs are so essential to the proof of the accuracy threshold theorem: (i) the number of errors these codes can correct grows exponentially with the number of layers of concatenation, and (ii) the time needed to do error correction is linear in the time it takes to error correct a code block in each layer.

4.1.2. Threshold Calculation—Principal Assumptions

It is perhaps not surprising to find that the specific value obtained for the accuracy threshold depends on the detailed assumptions made concerning the error model and the types of operations that can be applied to blocks of data and ancilla qubits. In this subsection we will state the principal assumptions on which the threshold calculations in Secs. 4.2 and 4.3 will be based. Secondary assumptions are also made in those sections when appropriate/convenient to simplify analysis of the calculational results.

Two assumptions were already stated in Sec. 4.1.1. We re-state them here so that all principal assumptions can be found in this subsection.

Assumption 4.1. *For a concatenated QECC, it is possible to measure the error syndrome for all code blocks in a given layer of concatenation simultaneously (viz. in parallel).*

Assumption 4.2. *Each layer of a concatenated QECC will be encoded using the same CSS code. In this section the [7,1,3] CSS code will be used for this purpose.*

The following three assumptions describe the nature of the errors that are assumed to occur.

Assumption 4.3. *The physical processes responsible for errors are such that independent errors form exclusive alternatives[35] and so the probabilities for independent errors add.*

Errors can occur on qubits as they wait around in storage registers, are operated on with quantum gates, and during measurements and state preparation.

Assumption 4.4. *Storage errors occur independently on different qubits with an error probability* **rate** p_{stor}.

Thus if a qubit is in storage for a time Δt, the probability that a storage error occurs is $p_{stor}\Delta t$. The error probability for quantum gates in the

Clifford group $N(\mathcal{G}_n)$ is p_g, and for the Toffoli gate is p_{Tof}. Note that the gate error probability is defined to *include* the storage error probability that accrues on qubits while they are being acted on by the gate.

Assumption 4.5. *Quantum gates can only produce errors on the qubits upon which they act. Qubits that are not the object of a gate operation do however accumulate storage errors while they wait for a gate to act on other qubits.*

The error probabilities for state preparation and measurement are denoted by p_{prep} and p_{meas}, respectively. The next assumption is a procedural one.

Assumption 4.6. *After any operation at code level-j, error correction is applied to all level-j code blocks. We denote the error probabilities for level-j Clifford group gates, Toffoli gates, and storage registers by $p_g^{(j)}$, $p_{Tof}^{(j)}$, and $p_{stor}^{(j)}$, respectively.*

The time required to implement an operation will depend on the code-level at which it is being applied.

Assumption 4.7. *The unit of time is chosen so that the time needed to apply an operation on a physical qubit at level-0 is 1.*

Clearly not all operations on level-0 qubits will take exactly one time-unit. It is being assumed that these operations will have application times that are of the same order of magnitude. This representative time then defines the unit of time, and we ignore the small differences in application times for the different gates. We denote the time it takes to apply a Toffoli gate at level-j by $t_{Tof}^{(j)}$, and the time to prepare an encoded state at level-j by $t_{prep}^{(j)}$. Our final assumption involves the times to do state preparation and measurement on physical qubits at level-0.

Assumption 4.8. *The time required to prepare the state of a level-0 qubit $t_{prep}^{(0)}$ is assumed to be negligible compared to the time to prepare a level-j encoded state for $j \geq 1$ and so it is set to zero: $t_{prep}^{(0)} = 0$. The time to measure the state of a level-0 qubit $t_{meas}^{(0)}$ is assumed to be of the same order of magnitude as the time to apply an operation to it and so $t_{meas}^{(0)} = 1$.*

As noted earlier, since we have restricted ourselves to a concatenated QECC built up from the [7,1,3] CSS code, all gates in the Clifford group can be applied at level-j by applying it transversally to qubits in the level-j code blocks. Because of the recursive structure of a concatenated QECC, this ultimately means that we apply the gate transversally to level-0 qubits. The time thus needed to implement a gate in the Clifford group at any code-level is simply the time it takes to apply the gate to a level-0 qubit and so $t_g^{(j)} = 1$.

Having listed our principal assumptions, we close this section with a statement of the Accuracy Threshold Theorem.

Theorem 4.1 (Accuracy Threshold Theorem). *A quantum computation of arbitrary duration can be done with arbitrarily small error probability in the presence of noise and using imperfect quantum gates if the following conditions hold:*

(1) the computational data is protected using a concatenated QECC;

(2) fault-tolerant preocedures are used for encoded quantum gates, error correction, and measurements; and

(3) storage registers and a universal set of unencoded quantum gates are available that have error probabilities P_e that are less than a threshold value P_a known as the accuracy threshold: $P_e < P_a$.

The proof of this theorem occupies the remainder of Sec. 4. The threshold calculations will use a concatenated QECC to encode the data, and all procedures will be fault-tolerant. Thus, the proof of the theorem will reduce to showing that the accuracy threshold takes on non-zero values for (i) a storage register, (ii) all gates in the Clifford group, and (iii) the Toffoli gate. The final value for the accuracy threshold for our set of assumptions will then be the smallest of these three values.

4.2. *Recursion Relation*

It proves convenient to think of a storage register as applying a quantum operation to the qubits stored in it. Ideally the operation is the identity. However, because of unwanted interactions with the environment, the operation actually applied will differ from the identity with a probability per unit time p_{stor}. In this section, when speaking of a generic quantum operation, we will have in mind either a gate in the Clifford group, a storage register "gate", or a Toffoli gate. Given an l-layered concatenated QECC, we refer to the error probabilities for level-l operations as effective error probabilities since they give the probability that a real error appears in the data qubit we are trying to protect. We refer to the error probability for level-0 operations as primitive error probabilities since they will be found to recursively determine the error probability for all higher-level operations.

To establish this connection, we derive a recursion-relation that connects the error probability $p_{op}^{(j)}$ for a level-j operation to the error probability of the level-$(j-1)$ operations that are used to apply it. The recursion-relation will then be used in Sec. 4.3 to determine the accuracy threshold for gates in the Clifford group, storage registers, and the Toffoli gate.

Recall that error correction is applied to every level-j code block after the application of a level-j quantum operation (Assumption 4.6). Let $p_{EC}^{(j)}$ denote the probability that error correction applied to a level-j code block produces an error on one of the level-$(j-1)$ qubits that belong to the block. By Assumption 4.2, we restrict ourselves to a concatenated QECC based on the [7,1,3] CSS code. Thus for a quantum operation to produce an error at level-j, at least two errors must occur at level-$(j-1)$ during the two-step procedure of quantum operation plus follow-up error correction. In the remainder of this section we focus on the most probable case where only two errors occur, and to simplify the analysis, assume that all code blocks acted on by a quantum operation begin the operation error-free.

Consider a level-j code block that has developed an error due to a level-j quantum operation. Two of the level-$(j-1)$ qubits belonging to the block must contain errors. There are $\binom{7}{2} = 21$ different ways in which two qubits can be chosen from a blcok of seven qubits to act as error sites; and for a given choice of qubits, the two errors can appear in one of three ways: (i) both during the quantum operation; (ii) one during the operation and the second during error correction; or (iii) both during error correction. For case (i), the quantum operation produces a level-j error if two of the level-$(j-1)$ operations used to apply it produce errors. This occurs[*1] with probability $(p_{op}^{(j-1)})^2$. For case (ii), the syndrome verification protocol of Sec. 3.2.3 requires that at most two syndrome extractions be carried out. Since an error has occurred during the quantum operation, the second error must occur during one of these two syndrome extractions. Let qubits A and B denote the two qubits on which the errors occur. The first error can occur on qubit A or B, and the second error can occur during the first or second syndrome extraction. Thus the total error probability for a case (ii) error is $4p_{op}^{(j-1)}p_{EC}^{(j)}$. For case (iii), both errors appear in the

---

[*1]This is true for gates in the Clifford group and storage registers. For a Toffoli gate, $p_{Tof}^{(j-1)}$ is *not* equal to the probability E that a level-$(j-1)$ qubit develops an error during a level-j Toffoli gate. We shall see (Sec. 4.3.2) that E is in fact a linear inhomogeneous function of $p_{Tof}^{(j-1)}$.

code block during error correction. Since the sydrome verification procedure requires at most two syndrome extraction operations, this can happen in two ways: either both errors occur during a single syndrome extraction; or when two syndrome extraction operations are applied, each produces an error. For the first scenario, the probability is $2\left(p_{EC}^{(j)}\right)^2$ since both errors occur during either the first or second syndrome extraction. For the second scenario, there are only two ways in which this scenario can occur. Either the first error appears on qubit A and the second on qubit B, or vice versa. The probability for the second scenario is thus $2\left(p_{EC}^{(j)}\right)^2$, and so the total probability for a case (iii) error is $4\left(p_{EC}^{(j)}\right)^2$. Putting together all these results we obtain the probability $p_{op}^{(j)}$ that a level-j quantum operation followed by error correction will produce a level-j error:

$$p_{op}^{(j)} = 21\left[\left(p_{op}^{(j-1)}\right)^2 + 4p_{op}^{(j-1)}p_{EC}^{(j)} + 4\left(p_{EC}^{(j)}\right)^2\right]. \tag{82}$$

To obtain our final expression for $p_{op}^{(j)}$ we must evaluate $p_{EC}^{(j)}$. We assume that error correction is being carried out using the fault-tolerant procedure presented in Sec. 3.2. Due to space limitations, we can only sketch out how that calculation is done here. The reader is referred to Ref. 1 for the details. There it is shown that $p_{EC}^{(j)}$ is the sum of the probabilities that (i) a phase error in the Shor state spreads to the data block during syndrome extraction, (ii) an error occurs during syndrome extraction, and (iii) a storage error appears during cat-state preparation and conversion, and during measurement of the error syndrome $S(e)$. One finds that

$$p_{EC}^{(j)} = 12p_g^{(j-1)} + \left(14 + t_{prep}^{(j-1)} + t_{meas}^{(j-1)}\right)p_{stor}^{(j-1)}. \tag{83}$$

Further calculation finds that $t_{meas}^{(j-1)} = 1$ and $t_{prep}^{(j-1)} = 32(j-1)$. Inserting these results into eq. (83) gives

$$p_{EC}^{(j)} = 12p_g^{(j-1)} + [15 + 32(j-1)]\,p_{stor}^{(j-1)}. \tag{84}$$

4.3. Accuracy Threshold

In this section we use eqs. (82) and (84) to determine the accuracy threshold for gates in the Clifford group and storage registers (Sec. 4.3.1), and the Toffoli gate (Sec. 4.3.2).

4.3.1. *Clifford Group Gates and Storage Registers*

Substituting eq. (84) into eq. (82) allows us to determine the recursion relation for the level-j error probabilities $p_g^{(j)}$ and $p_{stor}^{(j)}$ for gates in the Clifford group and for storage registers, respectively. One finds

$$p_g^{(j)} = 13125 \left(p_g^{(j-1)}\right)^2 + p_g^{(j-1)} p_{stor}^{(j-1)} \left[31500 + 67200(j-1)\right]$$
$$+ \left(p_{stor}^{(j-1)}\right)^2 \left[18900 + 80640(j-1) + 86016(j-1)^2\right], \quad (85)$$

and

$$p_{stor}^{(j)} = 12096 \left(p_g^{(j-1)}\right)^2 + p_g^{(j-1)} p_{stor}^{(j-1)} \left[31248 + 64512(j-1)\right]$$
$$+ \left(p_{stor}^{(j-1)}\right)^2 \left[20181 + 83328(j-1) + 86016(j-1)^2\right]. \quad (86)$$

We examine two cases that illustrate how to obtain the accuracy threshold from eqs. (85) and (86).

(i) Consider the case where level-0 storage errors occur with negligible probability: $p_{stor}^{(0)} = 0$. There is then no need to implement error correction on qubits that are not acted on by a quantum gate. For such a qubit, $p_{EC}^{(j)} = 0$ in eq. (82), and so for storage registers: $p_{stor}^{(j)} = 0$ for $0 \leq j \leq l$. Thus, for reliable storage registers, we only need to worry about the accuracy threshold for gates in the Clifford group $N(\mathcal{G}_n)$, and for the Toffoli gate. We use eq. (85) to find the accuracy threshold for gates in $N(\mathcal{G}_n)$, and will examine the Toffoli gate in the following subsection. Using $p_{stor}^{(j)} = 0$ in eq. (85) gives

$$p_g^{(j)} = 13125 \left(p_g^{(j-1)}\right)^2$$
$$= p_g^{(0)} \left(\frac{p_g^{(0)}}{p_{th}}\right)^{2^j - 1}, \quad (87)$$

where $p_{th} = 1/13125 = 7.62 \times 10^{-5}$ is the accuracy threshold for gates in the Clifford group when reliable storage registers are available.

(ii) Finally, we consider the case where ancilla blocks in the Shor state are always available when needed so that storage errors no longer accumulate over the time $t_{prep}^{(j-1)} + 6$ needed to prepare a level-$(j-1)$ Shor state (see Ref. 1). Then $p_{EC}^{(j)} = 12p_g^{(j-1)} + 9p_{stor}^{(j-1)}$. Using this in eq. (82) gives

$$p_g^{(j)} = 13125 \left(p_g^{(j-1)}\right)^2 + 18900 p_g^{(j-1)} p_{stor}^{(j-1)} + 6804 \left(p_{stor}^{(j-1)}\right)^2$$
$$p_{stor}^{(j)} = 12096 \left(p_g^{(j-1)}\right)^2 + 19152 p_g^{(j-1)} p_{stor}^{(j-1)} + 7581 \left(p_{stor}^{(j-1)}\right)^2.$$

Since the numerical coeeficients in these two equations are nearly equal, we approximate $p_g^{(j)} = p_{stor}^{(j)}$ for $j \geq 1$. Then

$$p_g^{(j)} = 38829 \left(p_g^{(j-1)} \right)^2$$

$$= p_g^{(0)} \left(\frac{p_g^{(0)}}{p_{th}} \right)^{2^j - 1},$$

where $p_{th} = 1/38829 = 2.58 \times 10^{-5}$.

4.3.2. Toffoli Gate

One task remains—to estimate the Toffoli gate accuracy threshold. In this subsection we will sketch out how this estimate is done. The full details can be found in Ref. 1. Recall how this gate is implemented (Sec. 3.3.7).

(1) We begin with six level-j qubits. Qubits 1-3 are prepared in the state

$$|A\rangle = (1/2) \left[|000\rangle + |010\rangle + |100\rangle + |111\rangle \right],$$

and qubits 4-6 in an arbitrary three-qubit state $|\psi\rangle$.
(2) Level-j CNOT gates are applied from qubits $1 \to 4$; $2 \to 5$; and $6 \to 3$.
(3) The encoded operators $Z_4^{(j)}$, $Z_5^{(j)}$, and $X_6^{(j)}$ are measured yielding outcomes m_4, m_5, and m_6, respectively.
(4) The operator $G_1^{(j)} = X_1^{(j)} U_{CNOT}^{23}$ is applied to qubits 1-3 when $m_4 = -1$. Similarly, $G_2^{(j)} = U_{CNOT}^{13} X_2^{(j)}$ $(G_3^{(j)} = U_{CP}^{12} Z_3^{(j)})$ is applied when m_5 (m_6) is -1. No operation is applied when a measurement outcome is $+1$.

In an effort to keep the notation from getting too cluttered, the level-j superscripts have been suppressed on $U_{CNOT}^{\mu\nu}$ and $U_{CP}^{\mu\nu}$. Notice that in Fig. 4 the final three CNOT gates, three measurements, and measurement-conditioned operations on qubits 1-3 implement steps 2-4 in the above Toffoli gate procedure. The remainder of the circuit in Fig. 4 places qubits 1-3 in the state $|A\rangle$ (to $\mathcal{O}(p)$) so that the complete circuit implements a level-j Toffoli gate.

To determine the Toffoli gate accuracy threshold we use eq. (82) to obtain a recursion relation for $p_{op}^{(j)} = p_{Tof}^{(j)}$. As input for that relation, we must find the probability E that a single error appears on a level-$(j-1)$ qubit during the level-j Toffoli gate. This probability is then identified with $p_{op}^{(j-1)}$ in eq. (82). To that end, since qubits 1-3 in Fig. 4 carry the data at the end of the circuit, it is necessary to determine the probability E_i that

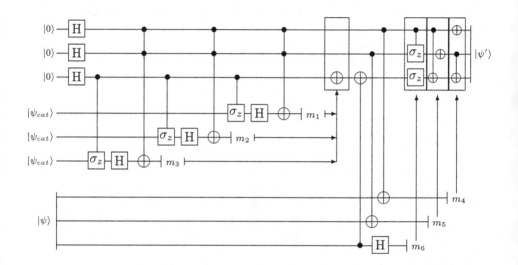

Fig. 4. Quantum circuit to implement a Toffoli gate U_T on level-j qubits: $|\psi'\rangle = U_T|\psi\rangle$. The top (bottom) three lines correspond to qubits 1–3 (4–6). Three ancilla blocks (denoted A_1–A_3 from top to bottom) are prepared in cat-states $|\psi_{cat}\rangle$, which are used to insure that (to $\mathcal{O}(p)$) qubits 1–3 are in the state $|A\rangle$ after the boxed NOT gate. Majority voting on the measurement outcomes $\{m_1, m_2, m_3\}$ determines whether the boxed NOT gate is applied. See text for further discussion. Based on Fig. 12, J. Preskill, *Proc. R. Soc. Lond. A* **454** (1998) 385. ©Royal Society, used with permission.

an error appears during the circuit on one of the level-$(j-1)$ qubits that constitute the ith qubit ($i = 1, 2, 3$). We then use the largest of the E_i to upper bound E. To simplify the analysis, one assumes that the initial state $|0\rangle$ for each of these qubits is error-free. In Ref. 1 we follow all the ways that an error can appear on qubits 1-3 to arrive at an expression for E. Due to space limitations we will not reproduce that discussion here. One finds that the error probabilities $\{E_i : i = 1, 2, 3\}$ upon completion of the circuit in Fig. 4 are:

$$E_1 = \left(6t_{cat} + 9t_{meas}^{(j-1)} + 12t_{Tof}^{(j-1)} + 41\right) p_{stor}^{(j-1)} + 9p_{cat} + 26p_g^{(j-1)} + 6p_{Tof}^{(j-1)}$$

$$E_2 = \left(6t_{cat} + 9t_{meas}^{(j-1)} + 12t_{Tof}^{(j-1)} + 41\right) p_{stor}^{(j-1)} + 9p_{cat} + 27p_g^{(j-1)} + 6p_{Tof}^{(j-1)}$$

$$E_3 = E_2. \tag{88}$$

Here t_{cat} and p_{cat} are the time to prepare a cat-state and the probability that the cat-state has a bit-flip error at the end of its preparation. From eq. (88) we see that $E = E_2$, and that E is not equal to $p_{Tof}^{(j-1)}$, but is instead a linear inhomogeneous function of this probability. It is E that must be plugged into eq. (82) for $p_{op}^{(j-1)}$ (see footnote on page 102). Doing so gives

$$p_{Tof}^{(j)} = 21\left[E^2 + 4Ep_{EC}^{(j)} + 4\left(p_{EC}^{(j)}\right)^2\right]. \tag{89}$$

We now work out the Toffoli gate accuracy threshold for the case where storage errors are negligible: $p_{stor}^{(0)} = 0$. As we saw in Secs. 4.2 and 4.3.1, this implies

$$p_{stor}^{(j-1)} = 0; \tag{90}$$

$$p_{EC}^{(j)} = 12p_g^{(j-1)}; \tag{91}$$

$$p_g^{(j)} = 13125\left(p_g^{(j-1)}\right)^2. \tag{92}$$

Eq. (92) reduces to $(j \to j - 1)$:

$$p_g^{(j-1)} = p_{th}\left(\frac{p_g^{(0)}}{p_{th}}\right)^{2^{(j-1)}}, \tag{93}$$

where $p_{th} = 1/13125 = 7.62 \times 10^{-5}$ is the accuracy threshold for gates in the Clifford group. Since storage errors are assumed to be negligible, it makes sense to verify the cat-state until we are sure its error probability is arbitrarily small (viz. $p_{cat} = 0$). Then eq. (89) becomes

$$p_{Tof}^{(j)} = 54621\left[p_g^{(j-1)}\right]^2 + 12852\left[p_g^{(j-1)}p_{Tof}^{(j-1)}\right] + 756\left[p_{Tof}^{(j-1)}\right]^2. \tag{94}$$

Now define $\epsilon = p_g^{(0)}/p_{th}$. Combining this with eqs. (93) and (94) gives

$$p_{Tof}^{(j)} = \left(3.17 \times 10^{-4}\right)\epsilon^{2^j} + \left[(0.98)\epsilon^{2^{(j-1)}} + 756p_{Tof}^{(j-1)}\right]p_{Tof}^{(j-1)}. \tag{95}$$

Suppose the level-0 Toffoli gate and Clifford group gates have comparable accuracy so that $p_{Tof}^{(0)} = p_g^{(0)}$, and that $\epsilon = p_g^{(0)}/p_{th} \lesssim 1$. Then $p_{Tof}^{(0)} = \epsilon p_{th}$, and from eq. (95) we have

$$p_{Tof}^{(1)} = \left(3.96 \times 10^{-4}\right)\epsilon^2; \tag{96}$$

$$p_{Tof}^{(2)} = \left(8.24 \times 10^{-4}\right)\epsilon^4; \tag{97}$$

$$p_{Tof}^{(3)} = \left(1.64 \times 10^{-3}\right)\epsilon^8. \tag{98}$$

Suppose that $p_{Tof}^{(3)} = p_{Tof}^{(2)}$. Then further concatenation will begin to reduce the effective error probability for the Toffoli gate. Equating eqs. (97) and

(98) gives $\epsilon^4 = 0.502$ and so $\epsilon = 0.84$. In this case, the Toffoli gate accuracy threshold is

$$P_a^{Tof} = \epsilon\, p_{th} = 6.4 \times 10^{-5}. \qquad (99)$$

Since the Toffoli gate threshold is smaller than the threshold p_{th} for gates in the Clifford group, it determines the accuracy threshold for reliable quantum computation for our set of assumptions. Section 4 has thus shown that if (i) computational data is protected using a sufficiently layered concatenated QECC; (ii) fault-tolerant procedures are used for quantum computation, error correction, and measurement; and (iii) storage registers and a universal set of unencoded quantum gates are available whose primitive error probabilities are less than $P_a = 6.4 \times 10^{-5}$, then the effective error probabilities for these registers and quantum gates can be made arbitrarily small. This in turn allows an arbitrarily long quantum computation to be done with arbitrarily small error probability, and so establishes the accuracy threshold theorem for the set of assumptions made earlier in this section. In principle, adaptation of the above arguments will allow the accuracy threshold theorem to be established for other sets of assumptions.

5. High-Fidelity Universal Quantum Gates Through Quantum Interference

As seen in Sec. 4, the accuracy threshold theorem establishes that reliable quantum computing is possible, even in the presence of noise and imperfect quantum gates. The theorem places demands on both the software and hardware used to carry out the quantum computation. From the software perspective, the computational data should be encoded using a concatenated QECC, and quantum computation, error correction, and measurement must be done fault-tolerantly. From the hardware perspective, a sufficiently reliable universal set of unencoded quantum gates must be available whose error probabilities all fall below a value known as the accuracy threshold. In this final section we describe an approach to making a universal set of unencoded quantum gates whose performance approaches the level demanded by the accuracy threshold theorem. The approach is based on controllable quantum interference effects that arise during a class of non-adiabatic rapid passage sweeps known as twisted rapid passage (TRP) which we introduce in Sec. 5.1. We then describe how the quantum dynamics generated by these sweeps is simulated, and the performance of the resulting quantum gates is determined (Sec. 5.2). Section 5.3 presents the universal gate set that will be produced using TRP, while Sec. 5.4 explains how group-symmetrized evolution can be used to improve the performance of the two-qubit gate in the TRP universal set. Best performance results for these gates are presented in Sec. 5.5, and we close with a discussion of these results in Sec. 5.6.

5.1. *Twisted Rapid Passage*

To begin, consider a single qubit interacting with an external control field $\mathbf{F}(t)$ via the Zeeman interaction

$$H_Z(t) = -\boldsymbol{\sigma} \cdot \mathbf{F}(t), \tag{100}$$

where $\{\sigma_i : i = x, y, z\}$ are the Pauli matrices. The sweeps we will be interested in are a generalization of those used in adiabatic rapid passage (ARP).[36] In ARP the field $\mathbf{F}(t)$ is inverted over a time T_0 such that $\mathbf{F}(t) = b\hat{\mathbf{x}} + at\hat{\mathbf{z}}$. The inversion time T_0 is large compared to the inverse Larmor frequency ω_0^{-1} (viz. adiabatic), though small compared to the thermal relaxation time τ_{th} (viz. rapid). ARP sweeps provide a highly precise method for inverting the Bloch vector $\mathbf{s}_i = \langle \boldsymbol{\sigma}_i \rangle$, although the price paid for this precision is an adiabatic inversion rate. We are interested in a type of rapid passage in which the control field $\mathbf{F}(t)$ is allowed to twist around in the x-y plane with time-varying azimuthal angle $\phi(t)$, while simultaneously undergoing inversion along the z-axis:

$$\mathbf{F}(t) = b \cos \phi(t)\hat{\mathbf{x}} + b \sin \phi(t)\hat{\mathbf{y}} + at\hat{\mathbf{z}}. \tag{101}$$

Here $-T_0/2 \le t \le T_0/2$. This class of rapid passage sweeps is known as twisted rapid passage (TRP). The first experimental realization of TRP in 1991 by Zwanziger et al.[37] carried out the inversion adiabatically with $\phi(t) = Bt^2$. Subsequently,[38] *non-adiabatic* TRP was studied with polynomial twist profile $\phi(t) = (2/n)Bt^n$, and controllable quantum interference effects were found to arise for $n \ge 3$. Zwanziger et al.[39] implemented non-adiabatic TRP with $n = 3, 4$ and observed the predicted interference effects. In the Zwanziger experiments, a TRP sweep is produced by sweeping the detector frequency $\dot{\phi}_{det}(t)$ linearly through resonance at the Larmor frequency ω_0: $\dot{\phi}_{det}(t) = \omega_0 + (2at)/\hbar$. The frequency of the rf-field $\dot{\phi}_{rf}(t)$ is also swept through resonance in such a way that $\dot{\phi}_{rf}(t) = \dot{\phi}_{det}(t) - \dot{\phi}(t)$, where $\phi(t) = (2/n)Bt^n$ is the TRP polynomial twist profile. Substituting the expression for $\dot{\phi}_{det}(t)$ into that for $\dot{\phi}_{rf}(t)$ gives

$$\dot{\phi}_{rf}(t) = \omega_0 + \frac{2at}{\hbar} - \dot{\phi}(t). \tag{102}$$

At resonance $\dot{\phi}_{rf}(t) = \omega_0$. Inserting this condition into eq. (102), it follows that at resonance

$$at - \frac{\hbar}{2}\dot{\phi}(t) = 0. \tag{103}$$

For polynomial twist $\phi(t) = (2/n)Bt^n$, eq. (103) has $n-1$ roots, though only real-valued roots correspond to resonance. Ref. 38 showed that for $n \geq 3$, multiple passes through resonance occur during a *single* TRP sweep: (i) for all n when $B > 0$; and (ii) for n odd when $B < 0$. We restrict ourselves to $B > 0$ in the remainder of this paper. In this case, the qubit passes through resonance at the times:

$$t = \left\{ \begin{array}{ll} 0, (a/\hbar B)^{\frac{1}{n-2}} & (n \text{ odd}) \\ 0, \pm (a/\hbar B)^{\frac{1}{n-2}} & (n \text{ even}) \end{array} \right. . \tag{104}$$

We see that the time separating the qubit resonances can be altered by variation of the sweep parameters B and a. Ref. 38 showed that these multiple resonances have a strong influence on the qubit transition probability. It was shown that qubit transitions could be significantly enhanced or suppressed by small variations of the sweep parameters, and hence of the time separating the resonances. Plots of transition probability versus time suggested that the multiple resonances were producing quantum interference effects that could be controlled by variation of the TRP sweep parameters. In Ref. 40 the qubit transition amplitude was calculated to all orders in the non-adiabatic coupling. The result found there can be re-expressed as the following diagrammatic series:

$$T_-(t) = \quad \text{⌐⌐} \quad + \quad \text{⌐⌐⌐} \quad + \quad \text{⌐⌐⌐⌐} \quad + \cdots . \tag{105}$$

Lower (upper) lines correspond to propagation in the negative (positive) energy level and the vertical lines correspond to transitions between the energy levels. The calculation sums the probability amplitudes for all interfering alternatives[35] that allow the qubit to end up in the positive energy level at time t given that it was initially in the negative energy level. As we have seen, varying the TRP sweep parameters varies the time separating the resonances. This in turn changes the value of each diagram in eq. (105), and thus alters the interference between alternatives in the quantum superposition. Similar diagrammatic series can be worked out for the remaining three combinations of final and initial states. It is the sensitivity of the individual alternatives/diagrams to the time separation of the resonances that allow TRP to manipulate this quantum interference. Zwanziger et al.[39]

observed these interference effects in the transition probability using liquid state NMR and found quantitative agreement between theory and experiment. It is the link between the TRP sweep parameters and this quantum interference that we believe makes it possible for TRP to drive highly accurate non-adiabatic one- and two-qubit gates. The results presented in Sec. 5.5 for the different gates in the TRP universal set $\mathcal{G}_u$ are found by numerical simulation of the Schrodinger equation. We next describe how these simulations are done and how gate performance is evaluated.

5.2. *Quantum Gate Dynamics and Performance*

A detailed presentation of our simulation and optimization protocols appear in Refs. 41 and 42. We will only give a brief sketch of that presentation here. As is well-known, the Schrodinger dynamics applies a unitary transformation U to an initial quantum state, with U generated by the system Hamiltonian $H(t)$. The Hamiltonian (see below) is assumed to contain terms that Zeeman-couple each qubit to the TRP control field $\mathbf{F}(t)$. Assigning values to the TRP sweep parameters (a,b,B,T_0) determines $H(t)$, which then determines the actual unitary transformation U_a applied. The task is to find sweep parameter values that produce a U_a that approximates a target gate U_t sufficiently closely that its error probability (defined below) satisfies $P_e < 10^{-4}$. In the following, the target gate U_t will be one of the gates in the TRP gate set $\mathcal{G}_u$ (Sec. 5.3). Since $\mathcal{G}_u$ contains only one- and two-qubit gates, our simulations only involve one- and two-qubit systems. For the one-qubit simulations, the Hamiltonian $H_1(t)$ is the Zeeman Hamiltonian (eq. (100)) introduced earlier. Ref. 42 showed that it can be written in the following dimensionless form:

$$\mathcal{H}_1(\tau) = \frac{1}{\lambda} \left[-\tau \sigma_z - \cos \phi_4(\tau) \sigma_x - \sin \phi_4(\tau) \sigma_y \right]. \tag{106}$$

Here: $\tau = (a/b)t$; $\lambda = \hbar a/b^2$; and for quartic twist, $\phi_4(\tau) = (\eta_4/2\lambda)\tau^4$ with $\eta_4 = \hbar B b^2/a^3$. We restrict our discussion to quartic twist in the remainder of Sec. 5, although other types of TRP have been considered in Refs. 38, 39, and 43. For the two-qubit simulations, the Hamiltonian $H_2(t)$ contains terms that Zeeman-couple each qubit to the TRP control field, and an Ising interaction term that couples the two qubits. Note that alternative two-qubit interactions can easily be considered, though we focus on the Ising interaction here. The energy-levels for the resulting Hamiltonian contain a resonance-frequency degeneracy that was found to spoil gate performance. Specifically, the resonance-frequency for transitions between the ground- and first-excited states $(E_1 \leftrightarrow E_2)$ is the same as that for transitions between the second- and third-excited states $(E_3 \leftrightarrow E_4)$. To remove this degeneracy a term $c_4|E_4(\tau)\rangle\langle E_4(\tau)|$ was added to $H_2(t)$. Combining all these remarks one arrives at the following dimensionless two-qubit Hamiltonian:[42]

$$
\begin{aligned}
\mathcal{H}_2(\tau) = {} & \left[-\left(d_1 + d_2\right)/2 + \tau/\lambda\right] \sigma_z^1 - (d_3/\lambda) \left[\cos\phi_4 \sigma_x^1 + \sin\phi_4 \sigma_y^1\right] \\
& + \left[-d_2/2 + \tau/\lambda\right] \sigma_z^2 - (1/\lambda) \left[\cos\phi_4 \sigma_x^2 + \sin\phi_4 \sigma_y^2\right] \\
& - (\pi d_4/2) \sigma_z^1 \sigma_z^2 + c_4|E_4(\tau)\rangle\langle E_4(\tau)|.
\end{aligned}
\tag{107}
$$

Here: (1) $b_i = \hbar\gamma_i B_{rf}/2$, $\omega_i = \gamma_i B_0$ and $i = 1, 2$; (2) $\tau = (a/b_2)t$, $\lambda = \hbar a/b_2^2$, and $\eta_4 = \hbar B b_2^2/a^3$; and (3) $d_1 = (\omega_1 - \omega_2)b_2/a$, $d_2 = (\Delta/a)b_2$, $d_3 = b_1/b_2$, and $d_4 = (J/a)b_2$, where Δ is a detuning parameter.[42]

The numerical simulation assigns values to the TRP sweep parameters and then integrates the Schrodinger equation to obtain the unitary transformation U_a produced by the sweep. To assess how closely U_a approximates the target gate U_t, it proves useful to introduce the positive operator $P = (U_a^\dagger - U_t^\dagger)(U_a - U_t)$. Given U_a, U_t, and an initial state $|\psi\rangle$, one can work out the error probability $P_e(\psi)$ for the TRP final-state $|\psi_a\rangle = U_a|\psi\rangle$, relative to the target final-state $|\psi_t\rangle = U_t|\psi\rangle$. The gate error probability P_e is defined to be the worst-case value of $P_e(\psi)$: $P_e \equiv \max_{|\psi\rangle} P_e(\psi)$. Ref. 41 showed that P_e satisfies the upper bound $P_e \leq Tr\,P$, where the RHS is the trace of the operator P introduced above. Once U_a is known, $Tr\,P$ is easily evaluated, and so it makes a convenient proxy for P_e, which is harder to calculate. To find TRP sweep parameter values that yield highly accurate non-adiabatic quantum gates, it proved necessary to combine the numerical simulations with function minimization algorithms that search for sweep parameter values that minimize the $Tr\,P$ upper bound.[44] The multi-dimensional downhill simplex method was used for the one-qubit gates in $\mathcal{G}_u$, while simulated annealing was used for the two-qubit gate (in $\mathcal{G}_u$). This produced the one-qubit gate results that will be presented in Sec. 5.5. However, for the two-qubit gate, simulated annealing was only able to find parameter values that gave $P_e \leq 1.27 \times 10^{-3}$ (see Ref. 42). To further improve the performance of the two-qubit gate, it proved necessary to incorporate the symmetrized evolution of Ref. 45 to obtain $P_e < 10^{-4}$. Symmetrized evolution will be described in Sec. 5.4. Before entering into that discussion, we introduce the gates that constitute the TRP universal gate set $\mathcal{G}_u$ in the following subsection.

5.3. *Twisted Rapid Passage Universal Gate Set*

As is well-known, an N-qubit quantum gate applies a fixed unitary transformation U to N-qubit states

$$|\psi_{out}\rangle = U|\psi_{in}\rangle. \tag{108}$$

We will be interested in the unitary transformation applied by the: (i) one-qubit Hadamard (U_H), phase (U_P), $\pi/8$ $(U_{\pi/8})$, and NOT (U_{NOT}) gates

$$U_H = \frac{1}{\sqrt{2}}\begin{pmatrix} 1 & 1 \\ 1 & -1 \end{pmatrix} \quad ; \quad U_P = \begin{pmatrix} 1 & 0 \\ 0 & i \end{pmatrix} \quad ; \tag{109}$$

$$U_{\pi/8} = \begin{pmatrix} 1 & 0 \\ 0 & e^{i\pi/4} \end{pmatrix} \quad ; \quad U_{NOT} = \begin{pmatrix} 0 & 1 \\ 1 & 0 \end{pmatrix} \quad , \tag{110}$$

and(ii) the two-qubit controlled-NOT (U_{CNOT}) and controlled-phase (U_{CP}) gates

$$U_{CNOT} = \begin{pmatrix} 1 & 0 & 0 & 0 \\ 0 & 1 & 0 & 0 \\ 0 & 0 & 0 & 1 \\ 0 & 0 & 1 & 0 \end{pmatrix} \quad ; \quad U_{CP} = \begin{pmatrix} 1 & 0 & 0 & 0 \\ 0 & 1 & 0 & 0 \\ 0 & 0 & 1 & 0 \\ 0 & 0 & 0 & -1 \end{pmatrix}. \tag{111}$$

The one- and two-qubit matrices appearing in eqs. (109)–(111) are in the representation spanned by the one- and two-qubit CB states $|i\rangle$ and $|ij\rangle$ which are, respectively, the eigenstates of σ_z and $\sigma_z^1 \otimes \sigma_z^2$.

Notice that the gates U_P and $U_{\pi/8}$ can be written as

$$U_P = e^{i\pi/4}U_{NOT}V_P \tag{112}$$

$$U_{\pi/8} = e^{i\pi/8}U_{NOT}V_{\pi/8}, \tag{113}$$

where

$$V_P = \begin{pmatrix} 0 & e^{i\pi/4} \\ e^{-i\pi/4} & 0 \end{pmatrix} \tag{114}$$

$$V_{\pi/8} = \begin{pmatrix} 0 & e^{i\pi/8} \\ e^{-i\pi/8} & 0 \end{pmatrix}, \tag{115}$$

and U_{NOT} is the NOT gate introduced above. Note also that U_{CNOT} can be written as

$$U_{CNOT} = \left(I^1 \otimes U_H^2\right)\left[\left(\sigma_z^1 \otimes I^2\right)V_{CP}\right]\left(I^1 \otimes U_H^2\right). \tag{116}$$

Here the superscript on a one-qubit gate labels the qubit on which the gate acts, and we have introduced the modified controlled-gate

$$V_{CP} = \begin{pmatrix} 1 & 0 & 0 & 0 \\ 0 & 1 & 0 & 0 \\ 0 & 0 & -1 & 0 \\ 0 & 0 & 0 & 1 \end{pmatrix}. \tag{117}$$

Finally, notice that the Pauli matrix σ_z^i can be implemented using the phase gate U_P^i: $\sigma_z^i = \left(U_P^i\right)^2$. It follows from eqs. (112), (113), and (116) that the set of gates $\{U_H, U_P, U_{\pi/8}, U_{CNOT}\}$ can be constructed using the set $\mathcal{G}_u = \{U_H, U_{NOT}, V_P, V_{\pi/8}, V_{CP}\}$. Since the first set of gates is universal,[46] so is the set $\mathcal{G}_u$. As will be seen below, TRP can be used to implement all gates in $\mathcal{G}_u$. For each gate we will present our best-case result and show how gate performance is altered by small variation of the parameters. One measure of gate performance is the gate error probability P_e which was introduced above and bounded by $Tr\,P$. A second measure is the gate fidelity $\mathcal{F}_n$ which is defined to be

$$\mathcal{F}_n \equiv \frac{1}{2^n}\mathrm{Re}\left[Tr\left(U_a^\dagger U_t\right)\right], \tag{118}$$

where n denotes the number of qubits acted on by the gate U_a, and U_t is the target gate. It is possible to relate our $Tr\,P$ upper bound to $\mathcal{F}_n$. Recalling that $P = \left(U_a - U_t\right)^\dagger \left(U_a - U_t\right)$, it is a simple matter to show[42] that

$$\mathcal{F}_n = 1 - \left(\frac{1}{2^{n+1}}\right) Tr\,P. \tag{119}$$

$Tr\,P$ thus yields the fidelity $\mathcal{F}_n$ and an upper bound on the gate error probability P_e. We now describe briefly how symmetrized evolution is incorporated into our simulations.

5.4. Group-Symmetrized Evolution

Ref. 45 introduced a unitary group-symmetrization procedure that yields an effective dynamics that is invariant under the action of a finite group $\mathcal{G}$. We incorporate this group-symmetrization into a TRP sweep by identifying the group $\mathcal{G}$ with a finite symmetry group of the target gate U_t, and then applying the procedure of Ref. 45 to filter out the $\mathcal{G}$-noninvariant part of the TRP dynamics. As the $\mathcal{G}$-noninvariant dynamics is manifestly bad dynamics relative to U_t, group-symmetrized TRP yields a better approximation to U_t. We begin by showing how the group-symmetrization procedure works, and then show how it can be incorporated into a TRP sweep.

Consider a quantum system Q with time-independent Hamiltonian H and Hilbert space $\mathcal{H}$. The problem is to provide Q with an effective dynamics that is invariant under a finite group $\mathcal{G}$, even when H itself is not $\mathcal{G}$-invariant. This symmetrized dynamics manifests as a $\mathcal{G}$-invariant effective propagator U that evolves the system over a time t. Let $\{\rho_i = \rho(g_i)\}$ be a unitary representation of $\mathcal{G}$ on $\mathcal{H}$, and let $|\mathcal{G}|$ denote the order of $\mathcal{G}$. The procedure begins by partitioning the time-interval $(0, t)$ into N subintervals of duration $\Delta t_N = t/N$, and then further partitioning each subinterval into $|\mathcal{G}|$ smaller intervals of duration $\delta t_N = \Delta t_N/|\mathcal{G}|$. Let $\delta U_N = \exp\left[-(i/\hbar)\,\delta t_N H\right]$ denote the H-generated propagator for a time-interval δt_N, and assume that the time to apply each $\rho_i \in \mathcal{G}$ is negligible compared to δt_N (bang-bang limit[47]). In each subinterval, the following sequence of transformations is applied:

$$U(\Delta t_N) = \prod_{i=1}^{|\mathcal{G}|} \rho_i^{\dagger} \delta U_N \rho_i. \qquad (120)$$

Ref. 45 showed that: (i) $U(\Delta t_N) \to \exp\left[-(i/\hbar)\Delta t_N \tilde{H}\right]$ as $N \to \infty$, where $\tilde{H} = (1/|\mathcal{G}|)\sum_{i=1}^{|\mathcal{G}|} \rho_i^{\dagger} H \rho_i$; (ii) $\tilde{H}$ is $\mathcal{G}$-invariant ($[\tilde{H}, \rho_i] = 0$ for all $\rho_i \in \mathcal{G}$); and (iii) the propagator $\tilde{U}$ over $(0, t)$ is $\tilde{U} = \exp\left[-(i/\hbar)t\tilde{H}\right]$, which is $\mathcal{G}$-invariant due to the $\mathcal{G}$-invariance of $\tilde{H}$. The end result is an effective propagator $\tilde{U}$ that is $\mathcal{G}$-invariant as desired.

This procedure can be generalized to allow for a time-dependent Hamiltonian $H(t)$. To do this, the time interval $(0, t)$ must be divided into sufficiently small subintervals that $H(t)$ is effectively constant in each. Within each subinterval, the above time-independent argument is applied, yielding a $\mathcal{G}$-symmetrized propagator for that subinterval. Combining the effective propagators for each of the subintervals then gives the full propagator

$$\tilde{U} = T\left[\exp\left(-i/\hbar \int_0^t d\tau \tilde{H}(\tau)\right)\right], \qquad (121)$$

where T indicates a time-ordered exponential, and

$$\tilde{H}(t) = (1/|\mathcal{G}|)\sum_{i=1}^{|\mathcal{G}|} \rho_i^{\dagger} H(t)\rho_i. \qquad (122)$$

For our two-qubit simulations, the target gate is V_{CP} which can be written as

$$V_{CP} = \frac{1}{2}\left[\left(I^1 + \sigma_z^1\right)I^2 - \left(I^1 - \sigma_z^1\right)\sigma_z^2\right] \qquad (123)$$

which is invariant under the group $\mathcal{G} = \{I, \sigma_z^1, \sigma_z^2, \sigma_z^1 \sigma_z^2\}$. Thus $|\mathcal{G}| = 4$, and we set $\rho_1 = I, \cdots, \rho_4 = \sigma_z^1 \sigma_z^2$. Switching over to dimensionless time τ, we partition the sweep time-interval $(-\tau_0/2, \tau_0/2)$ into sufficiently small subintervals that our two-qubit Hamiltonian $\mathcal{H}_2(\tau)$ is effectively constant within each. We then apply the time-independent symmetrization procedure to each subinterval with the V_{CP} symmetry group acting as $\mathcal{G}$. Combining the effective propagators for each of the subintervals as above gives the $\mathcal{G}$-symmetrized propagator for the full TRP sweep $\tilde{U} = T \left[\exp \left(-i/\hbar \int_{-\tau_0/2}^{\tau_0/2} d\tau \tilde{H}(\tau) \right) \right]$, with $\tilde{H}(\tau) = (1/4) \sum_{i=1}^{4} \rho_i^\dagger \mathcal{H}_2(\tau) \rho_i$. We shall see that $\mathcal{G}$-symmetrized TRP yields an approximation to V_{CP} with $P_e < 10^{-4}$.

5.5. *Gate Results*

All results presented in this subsection are for quartic TRP which has twist profile

$$\phi(\tau) = \frac{1}{2} \left(\frac{\eta_4}{\lambda} \right) \tau^4. \tag{124}$$

Here τ, λ, and η_4 are the dimensionless versions of time t, inversion rate a, and twist strength B. Their definitions appear immediately following eq. (106). All simulations were done with $\lambda > 1$ corresponding to non-adiabatic inversion, and the dimensionless inversion time $\tau_0 = aT_0/b$ was fixed at 80.000 for the one-qubit simulations, and at 120.00 for the two-qubit simulations.

The translation key connecting the simulation parameters to the experimental sweep parameters used in the Zwanziger experiments[37,39] is given in the Appendix of Ref. 38. We re-write the formulas for quartic twist here for convenience. Note that Zwanziger's symbol B is here replaced by $\mathcal{B}$ to avoid confusion with our use of B to denote the twist strength. First we give the formulas connecting our parameters (a, b, B, T_0) to the Zwanziger parameters $(\omega_1, A, \mathcal{B}, T_0)$:

$$\omega_1 = \frac{2b}{\hbar} \tag{125}$$

$$A = \frac{aT_0}{\hbar} \tag{126}$$

$$\mathcal{B} = \frac{BT_0^4}{2}, \tag{127}$$

where the inversion time T_0 is common to both parameter sets. The formulas linking the dimensionless sweep parameters (λ, η_4) to the Zwanziger parameters $(\omega_1, A, \mathcal{B}, T_0)$ are:

$$\lambda = \frac{4A}{\omega_1^2 T_0} \tag{128}$$

$$\eta_4 = \frac{\mathcal{B}\omega_1^2}{2A^3 T_0}. \tag{129}$$

In the experiments of Ref. 39: $\omega_1 = 393\,\mathrm{Hz}$; $T_0 = 41.00\,\mathrm{ms}$; $A = 50\,000\,\mathrm{Hz}$; and $\mathcal{B}$ was calculated from eq. (129) with η_4 varying over the range $[4.50, 4.70] \times 10^{-4}$. We are now ready to present our results.[41,42,48]

Hadamard Gate

The sweep parameters $\lambda = 5.8511$ and $\eta_4 = 2.9280 \times 10^{-4}$ produced the gate U_a whose real and imaginary parts are:

$$Re(U_a) = \begin{pmatrix} 0.708581 & 0.705629 \\ 0.705629 & -0.708581 \end{pmatrix} \tag{130}$$

$$Im(U_a) = \begin{pmatrix} 0.380321 \times 10^{-9} & -0.144317 \times 10^{-4} \\ 0.144317 \times 10^{-4} & 0.420313 \times 10^{-9} \end{pmatrix}. \tag{131}$$

For comparison, the real and imaginary parts of the target Hadamard gate $U_t = U_H$ are:

$$Re(U_H) = \begin{pmatrix} 0.707107 & 0.707107 \\ 0.707107 & -0.707107 \end{pmatrix} \tag{132}$$

$$Im(U_H) = \begin{pmatrix} 0 & 0 \\ 0 & 0 \end{pmatrix}. \tag{133}$$

From U_a and U_H we find $Tr\,P = 8.82 \times 10^{-6}$. This yields a gate fidelity of $\mathcal{F}_H = 0.999998$, and the gate error probability satisfies $P_e \leq 8.82 \times 10^{-6}$. Table 1 shows how gate performance varies when the sweep parameters are

Table 1. Variation of $Tr\,P$ for the Hadamard gate when the TRP sweep parameters are altered slightly from their best performance values. The columns to the left of center have $\eta_4 = 2.9280 \times 10^{-4}$ and those to the right have $\lambda = 5.8511$.

η_4	λ	$Tr\,P$	λ	η_4	$Tr\,P$
2.9280×10^{-4}	5.8510	7.22×10^{-5}	5.8511	2.9279×10^{-4}	7.03×10^{-4}
	5.8511	8.82×10^{-6}		2.9280×10^{-4}	8.82×10^{-6}
	5.8512	1.84×10^{-5}		2.9281×10^{-4}	6.14×10^{-4}

altered slightly. Of the two sweep parameters, η_4 variation is seen to have the largest impact on gate performance. This will turn out to be true for the other one-qubit gates as well. Although TRP can produce a Hadamard gate whose error probability falls below the accuracy threshold $P_a \sim 10^{-4}$, it is clear that the sweep parameters must be controlled to 5 significant figures to achieve this level of performance. See Sec. 5.6 for further discussion of this point.

V_P Gate

The target gate here is V_P. The sweep parameters $\lambda = 5.9750$ and $\eta_4 = 3.8060 \times 10^{-4}$ produce the gate U_a:

$$Re(U_a) = \begin{pmatrix} -0.627432 \times 10^{-2} & 0.706181 \\ 0.706181 & 0.627432 \times 10^{-2} \end{pmatrix} \tag{134}$$

$$Im(U_a) = \begin{pmatrix} -0.284521 \times 10^{-10} & 0.708004 \\ -0.708004 & 0.694222 \times 10^{-11} \end{pmatrix}. \tag{135}$$

The real and imaginary parts of the target gate V_P are:

$$Re(V_P) = \begin{pmatrix} 0 & 0.707107 \\ 0.707107 & 0 \end{pmatrix} \tag{136}$$

$$Im(V_P) = \begin{pmatrix} 0 & 0.707107 \\ -0.707107 & 0 \end{pmatrix}. \tag{137}$$

From U_a and V_P we find $Tr\,P = 8.20 \times 10^{-5}$, which yields a gate fidelity $\mathcal{F}_{V_P} = 0.999980$, and $P_e \leq 8.20 \times 10^{-5}$. Table 2 shows how $Tr\,P$ varies

Table 2. Variation of $Tr\,P$ for the V_P gate when the TRP sweep parameters are altered slightly from their best performance values. The columns to the left of center have $\eta_4 = 3.8060 \times 10^{-4}$ and those to the right have $\lambda = 5.9750$.

η_4	λ	$Tr\,P$	λ	η_4	$Tr\,P$
3.8060×10^{-4}	5.9749	1.56×10^{-4}	5.9750	3.8059×10^{-4}	2.29×10^{-3}
	5.9750	8.20×10^{-5}		3.8060×10^{-4}	8.20×10^{-5}
	5.9751	1.43×10^{-4}		3.8061×10^{-4}	1.88×10^{-3}

when η_4 and λ are varied slightly. Again gate performance is most sensitive to variation of η_4, and the sweep parameters must be controlled to high precision for performance to surpass the accuracy threshold (see Sec. 5.6).

$V_{\pi/8}$ *Gate*

The target gate this time is $V_{\pi/8}$. For $\lambda = 6.0150$ and $\eta_4 = 8.1464 \times 10^{-4}$ TRP produced the gate U_a:

$$Re(U_a) = \begin{pmatrix} 0.101927 \times 10^{-2} & 0.925307 \\ 0.925307 & -0.101927 \times 10^{-2} \end{pmatrix} \tag{138}$$

$$Im(U_a) = \begin{pmatrix} -0.960223 \times 10^{-10} & 0.379218 \\ -0.379218 & 0.184961 \times 10^{-10} \end{pmatrix}. \tag{139}$$

The target gate $V_{\pi/8}$ is:

$$Re(V_{\pi/8}) = \begin{pmatrix} 0 & 0.923880 \\ 0.923880 & 0 \end{pmatrix} \tag{140}$$

$$Im(V_{\pi/8}) = \begin{pmatrix} 0 & 0.382683 \\ -0.382683 & 0 \end{pmatrix}. \tag{141}$$

These matrices give $Tr\,P = 3.03 \times 10^{-5}$, $\mathcal{F}_{V_{\pi/8}} = 0.999992$, and $P_e \leq 3.03 \times 10^{-5}$. Table 3 shows how gate performance varies when the sweep

Table 3. Variation of $Tr\,P$ for the $V_{\pi/8}$ gate when the TRP sweep parameters are altered slightly from their best performance values. The columns to the left of center have $\eta_4 = 8.1464 \times 10^{-4}$ and those to the right have $\lambda = 6.0150$.

η_4	λ	$Tr\,P$	λ	η_4	$Tr\,P$
8.1464×10^{-4}	6.0149	1.30×10^{-3}	6.0150	8.1463×10^{-4}	1.77×10^{-3}
	6.0150	3.03×10^{-5}		8.1464×10^{-4}	3.03×10^{-5}
	6.0151	2.18×10^{-3}		8.1465×10^{-4}	2.77×10^{-3}

parameters are altered slightly. As with the previous two gates, performance is most sensitive to variation of η_4, and sweep parameters must be controllable to high precision (see Sec. 5.6).

NOT Gate

For parameters $\lambda = 7.3205$ and $\eta_4 = 2.9277 \times 10^{-4}$ TRP produced the gate U_a:

$$Re(U_a) = \begin{pmatrix} 0.235039 \times 10^{-2} & 0.999997 \\ 0.999997 & -0.235039 \times 10^{-2} \end{pmatrix} \tag{142}$$

$$Im(U_a) = \begin{pmatrix} -0.323648 \times 10^{-10} & -0.115151 \times 10^{-4} \\ 0.115150 \times 10^{-4} & 0.271006 \times 10^{-10} \end{pmatrix}. \tag{143}$$

For comparison, U_{NOT} is:

$$Re(U_{NOT}) = \begin{pmatrix} 0 & 1 \\ 1 & 0 \end{pmatrix} \qquad (144)$$

$$Im(U_{NOT}) = \begin{pmatrix} 0 & 0 \\ 0 & 0 \end{pmatrix} \quad . \qquad (145)$$

These matrices yield $Tr\, P = 1.10 \times 10^{-5}$, $\mathcal{F}_{NOT} = 0.999997$, and $P_e \leq 1.10 \times 10^{-5}$. Table 4 shows how $Tr\, P$ varies with small variation of the

Table 4. Variation of $Tr\, P$ for the NOT gate when the TRP sweep parameters are altered slightly from their best performance values. The columns to the left of center have $\eta_4 = 2.9277 \times 10^{-4}$ and those to the right have $\lambda = 7.3205$.

η_4	λ	$Tr\, P$	λ	η_4	$Tr\, P$
2.9277×10^{-4}	7.3204	1.12×10^{-5}	7.3205	2.9276×10^{-4}	1.23×10^{-3}
	7.3205	1.10×10^{-5}		2.9277×10^{-4}	1.10×10^{-5}
	7.3206	1.22×10^{-5}		2.9278×10^{-4}	1.23×10^{-3}

sweep parameters. As with the other gates, performance is most sensitive to variation in η_4, and sweep parameters must be controllable to 5 significant figures for the gate error probability P_e to fall below the accuracy threshold $P_a \sim 10^{-4}$ (see Sec. 5.6).

V_{CP} Gate

We complete the universal gate set $\mathcal{G}_u$ by presenting our simulation results for the $\mathcal{G}$-symmetrized TRP implementation of the modified controlled-phase gate V_{CP}. The target gate here is V_{CP} whose real part is the matrix $diag(1, 1, -1, 1)$ and whose imaginary part vanishes. TRP implementation of V_{CP} *without* symmetrized evolution was reported in Ref. 42. The result presented there is superceded by the $\mathcal{G}$-symmetrized TRP result presented below.[48] For purposes of later discussion, note that the parameters appearing in $\mathcal{H}_2(\tau)$ fall into two sets. The first set consists of the TRP sweep parameters $(\lambda, \eta_4, \tau_0)$, while the second set $(c_4, d_1, \ldots, d_4)$ consists of parameters for degeneracy-breaking, detuning, and coupling. We partitioned the TRP sweep into $N_{seq} = 2500$ pulse sequences, with each sequence based on the 4-element symmetry group for V_{CP} introduced earlier. The optimized parameter values $\lambda = 5.04$, $\eta_4 = 3.0 \times 10^{-4}$, $\tau_0 = 120.00$, $c_4 = 2.173$, $d_1 = 99.3$, $d_2 = 0.0$, $d_3 = -0.41$, and $d_4 = 0.8347$ produced the following two-qubit gate U_a:

$$Re(U_a) = \begin{pmatrix} 0.999\,998 & -0.000\,003 & -0.000\,015 & -0.000\,014 \\ 0.000\,003 & 0.999\,997 & 0.000\,036 & 0.000\,261 \\ -0.000\,015 & 0.000\,034 & -0.999\,980 & -0.003\,818 \\ -0.000\,014 & -0.000\,257 & -0.003\,838 & 0.999\,981 \end{pmatrix} \quad (146)$$

$$Im(U_a) = \begin{pmatrix} -0.002\,151 & 0.000\,003 & -0.000\,010 & -0.000\,073 \\ -0.000\,003 & -0.002\,180 & 0.000\,140 & -0.000\,325 \\ 0.000\,010 & -0.001\,140 & 0.001\,702 & 0.004\,534 \\ -0.000\,073 & -0.000\,328 & -0.004\,521 & -0.001\,778 \end{pmatrix} . \quad (147)$$

From U_a and V_{CP} we find that $Tr\,P = 8.87 \times 10^{-5}$; gate fidelity $\mathcal{F}_{CP} = 0.9999\,89$; and $P_e \leq 8.87 \times 10^{-5}$. We see that by adding symmetrized evolution to a TRP sweep we obtain an approximation to V_{CP} with $P_e < 10^{-4}$. Table 5 shows how $Tr\,P$ varies when either c_4 or d_4 is varied slightly, with

Table 5. Variation of $Tr P$ for V_{CP} when the parameters c_4 and d_4 are altered slightly from their best performance values. The columns to the left have $d_4 = 0.8347$ and those to the right have $c_4 = 2.173$. All other parameter values are as given in text.

d_4	c_4	$Tr P$	c_4	d_4	$Tr P$
0.8347	2.172	6.79×10^{-3}	2.173	0.8346	1.52×10^{-3}
	2.173	8.87×10^{-5}		0.8347	8.87×10^{-5}
	2.174	7.73×10^{-3}		0.8348	1.52×10^{-3}

all other parameters held fixed. Gate performance is found to be most sensitive to these two parameters. Note that these parameters only need to be controlled to 4 significant figures as a result of the $\mathcal{G}$-symmetrized evolution (see Sec. 5.6 for further discussion).

5.6. *Discussion*

We have presented in Sec. 5.5 simulation results which suggest that TRP sweeps should be capable of implementing a universal set of quantum gates $\mathcal{G}_u$ that operate non-adiabatically and with gate error probabilities satisfying $P_e < 10^{-4}$. To achieve this high level of performance in our current formulation of TRP, some of the TRP parameters must be controllable to high precision. For the one-qubit gates,[41] the critical parameter is η_4 which must be controlled to 5 significant figures to achieve best gate performance.

For the modified controlled-phase gate V_{CP}, the critical parameters are *not* the TRP sweep parameters. Instead, for V_{CP} *without* symmetrized evolution,[42] the critical parameters are c_4, d_1, and d_4 which also require 5 significant figure precision. However, when symmetrized evolution is added, not only is TRP able to make V_{CP} with $P_e < 10^{-4}$, but *gate robustness is also improved*. Specifically, d_1 ceases to be a critical parameter, and c_4 and d_4 now only need to be controllable to 4 significant figures (see Table 5). Thus adding symmetrized evolution improves both the accuracy and robustness of the TRP approximation to V_{CP}. Unfortunately, symmetrized evolution cannot be used to improve the robustness of the one-qubit gates. It is possible to show that if $U_t = \mathbf{a} \cdot \boldsymbol{\sigma}$, the only one-qubit unitary operators that commute with U_t are the identity and a multiple of U_t. Thus the only symmetry group available that does not include U_t is the trivial group composed of solely the identity. Some other means will have to be found to improve the robustness of the one-qubit gates in $\mathcal{G}_u$. Enhancing TRP gate robustness is now the main challenge that must be overcome if these sweeps are to become a viable means of universal control of a quantum computer.

Refs. 38, 41, and 42 have shown how TRP sweeps can be applied to NMR, atomic, and superconducting qubits. It should also be possible to apply these sweeps to spin-based qubits in quantum dots using magnetic fields since the same Zeeman-coupling acts here as with NMR qubits. Our focus in the near future will be to develop a better analytical understanding of TRP, with the aim being to find a way to make $Tr\,P$ for the one-qubit gates a more slowly varying function of the TRP parameters. Note that the precision needed for V_{CP} (with symmetrized evolution) is probably within the reach of current technology. We would also like to develop a theory for the optimum twist profile $\phi(t)$ for a given target gate.

Acknowledgments

We would like to thank Professor Mikio Nakahara for the opportunity to present the contents of this Chapter as a series of lectures at the *Symposium on Decoherence Suppression in Quantum Systems* held at the Oxford Kobe Institute on September 7-10, 2008. We also thank Franco Nori, RIKEN, and CREST for making our visit to RIKEN possible, and to RIKEN for access to the RIKEN Super Combined Cluster on which the simulations incorporating symmetrized evolution were done. Finally, one of us (F.G.) thanks T. Howell III for continued support.

References

1. F. Gaitan: *Quantum Error Correction and Fault-Tolerant Quantum Computing* (Taylor & Francis/CRC Press, Boca Raton, FL, 2008).
2. F.J. McWilliams and N.J.A. Sloane: *The Theory of Error Correcting Codes* (North-Holland, New York, 1977).
3. E.Knill and R.Laflamme: Phys. Rev. A **55** (1997) 900.
4. K. Kraus: *States, Effects, and Operations: Fundamental Notions of Quantum Theory*, Lecture Notes in Physics, Vol. 190 (Springer-Verlag, Berlin, 1983).
5. M.A. Nielsen and I.L. Chuang: *Quantum Computation and Quantum Information* (Cambridge University Press, London, 2000).
6. D. Gottesman: Phys. Rev. A **54** (1996) 1862.
7. A.R. Calderbank et al.: Phys. Rev. Lett. **78** (1997) 405.
8. A.R. Calderbank et al.: IEEE Trans. Inf. Theor. **44** (1998) 1369.
9. A.R. Calderbank and P.W. Shor: Phys. Rev. A **54** (1996) 1098.
10. A.M. Steane: Proc. R. Soc. Lond. A **452** (1996) 2551.
11. E. Knill and R. Laflamme: download at http://arXiv.org/abs/quant-ph/9608012, 1996.
12. D. Aharonov and M. Ben-Or: in *Proceedings of the Twenty-Ninth ACM Symposium on the Theory of Computing*, 1997, pp. 176-188.
13. P.W. Shor: Phys. Rev. A **52** (1995) 2493.
14. For an review of quantum circuit theory, see Ref. 5.
15. M.B. Plenio, V. Vedral and P.L. Knight: Phys. Rev. A **55** (1997) 4593.
16. P.W. Shor: in *Proceedings, 37th Annual Symposium on Fundamentals of Computer Science* (IEEE Press, Los Alamitos, CA, 1996), pp. 56-65.
17. J. Preskill: in *Introduction to Quantum Computation and Information*, H.-K. Lo, S. Popescu and T. Spiller, eds. (World Scientific, Singapore, 1998).
18. A.M. Steane: Phys. Rev. Lett. **78** (1997) 2252.
19. A.M. Steane: Nature **399** (1999) 124.
20. A.Y. Kitaev: Russ. Math. Surv. **52** (1997) 1191.
21. A.R. Calderbank et al.: Phys. Rev. Lett. **78** (1997) 405.
22. A.R. Calderbank et al.: IEEE Trans. Inf. Theor. **44** (1998) 1369.
23. C.H. Bennett et al.: Phys. Rev. A **54** (1996) 3824.
24. D. Gottesman: Phys. Rev. A **57** (1998) 127.
25. D. Gottesman and I. Chuang: Nature **402** (1999) 390.
26. X. Zhou et al.: Phys. Rev. A **62** (2000) 052316.
27. E. Knill and R. Laflamme: available at arXiv.org/quant-ph/960812, 1996.
28. E. Knill, R. Laflamme and W.H. Zurek: Science **279** (1998) 342.
29. E. Knill, R. Laflamme and W.H. Zurek: Proc. R. Soc. Lond. A **454** (1998) 365.
30. C. Zalka: available at arXiv.org/quant-ph/9612028, 1996.
31. D. Gottesman: PhD thesis, California Institute of Technology, Pasadena, CA (1997).
32. J. Preskill: Proc. R. Soc. Lond. A **454** (1998) 385.
33. A.Y. Kitaev: Russ. Math. Surv. **52** (1997) 1191.
34. A. Y. Kitaev: in *Quantum Communication, Computing, and Measurement* (Plenum Press, New York, 1997).

35. R.P. Feynman and A.R. Hibbs: *Quantum Mechanics and Path Integrals* (McGraw-Hill, New York, 1965).
36. A. Abragam: *Principles of Nuclear Magnetism* (Oxford University Press, New York, 1961).
37. J.W. Zwanziger, S.P. Rucker and G.C. Chingas: Phys. Rev. A **43** (1991) 3232.
38. F. Gaitan: Phys. Rev. A **68** (2003) 052314.
39. J.W. Zwanziger, U. Werner-Zwanziger and F. Gaitan, Chem. Phys. Lett. **375** (2003) 429.
40. F. Gaitan: J. Mag. Resonance **139** (1999) 152, see eq. (14).
41. R. Li, M. Hoover and F. Gaitan: Quant. Info. Comp. **7** (2007) 594.
42. R. Li, M. Hoover and F. Gaitan: Quant. Info. Comp. **9** (2009) 290.
43. R. Li and F. Gaitan: Optics and Spectroscopy **99** (2005) 257.
44. W.H. Press et al.: *Numerical Recipes* (Cambridge University Press, New York, 1992).
45. P. Zanardi: Phys. Lett. A **258** (1999) 77.
46. P.O. Boykin et al.: in *Proc. 40th Ann. Symp. on Found. Comp. Sc.* (1999), p. 486.
47. L. Viola and S. Lloyd: Phys. Rev. A **58** (1998) 2733.
48. R. Li and F. Gaitan: available at arXiv.org:0810.0741, 2008.

COMPOSITE PULSES AS GEOMETRIC QUANTUM GATES

YUKIHIRO OTA[*,§] and YASUSHI KONDO[*,†,¶]

*Interdisciplinary Graduate School of Science and Engineering,
Kinki University University,
3-4-1 Kowakae Higashi-Osaka, 577-8502, Japan
†Department of Physics, Kinki University University,
3-4-1 Kowakae Higashi-Osaka, 577-8502, Japan
E-mail: §yota@alice.math.kindai.ac.jp
¶kondo@phys.kindai.ac.jp

We show that some composite pulses are regarded as a non-adiabatic geometric quantum gate. We propose a model to discuss the effect of fluctuations on the non-adiabatic geometric quantum gates. The current model is simple, and easy to see the physical origin of the fluctuation. We classify a type of the noise for which the non-adiabatic geometric quantum gates are robust.

Keywords: Composite Pulses; Geometric Phase

1. Introduction

Reliable and stable control schemes of a quantum system are necessary to realize quantum information processing (QIP). Application of geometric phases QIP is interesting and promising. Various theoretical end experimental results have been reported.[1-7] Geometric phases are expected to be robust against noise because they are independent of details of dynamics. However, the robustness of a geometric quantum gate (GQG), which is a quantum gate only with geometric phases, is still open question. Theoretical attempts to verify it have been reported.[8-15] Blais and Tremblay[9] claimed that no advantage of the GQGs existed compared to corresponding quantum gates with dynamical phases, while Zhu and Zanardi[12] showed that their non-adiabatic GQGs are robust against the fluctuation of control parameters.

In this lecture note, we show some composite pulses[16-19] are regared as non-adiabatic GQGs.[15] Composite pulses are well-known techniques in the field of NMR to accomplish stable manipulation of spin dynamics in the presence of experimental imperfection (e.g., off-resonance effects). It is interesting that geometrical phases exist in the background of some composite pulses. Namely, we can obtain another clue to understand the physics of composite pulses. In addition, we propose a model to discuss the effect of fluctuations on the non-adiabatic GQGs. We classify a type of the noise for which the non-adiabatic GQGs are robust.

This lecture note is organized as follows. We briefly explain the composite pulses in Sec. 2. We review the definition of the geometric phase was introduced Aharanov and Anandan[20] and the geometric phases in more general settings (e.g., in non-unitary time evolution) in Sec. 3. In Sec. 4, we explain how to construct one-qubit non-adiabatic GQGs and how to eliminate the contribution of dynamical phases. One of our main results are shown in Sec. 5; a kind of the composite pulse can be regarded as a non-adiabatic GQG. Then, we propose a simple and practical model for examining the effect of fluctuations on non-adiabatic GQGs in Sec. 6. Section 7 is devoted to the summary.

2. Composite Pulses

2.1. *An example: population inversion*

First, we explain a simple composite pulse and its effectiveness. Let us consider the population inversion. In this operation, the value of the z-component of the magnetization M_z is inverted (i.e., $M_z \to -M_z$). The essential operation to do this is the rotation around y-axis with the angle $\pi/2$, $e^{-i\pi\sigma_y/2}$. In a NMR system, one can design such a operator in a rotating frame, in which the angular frequency of the radiofrequency (rf) pulse is equal to the Larmour frequency of a nuclear spin.[21] However, experimental imperfections always exist. The resonance condition might be slightly violated (i.e., the off-resonance effect). The shape and the phase of the rf pulses could be disturbed. Accordingly, the implementation by such a single pulse doesn't work well in real experiments. A compensated population inversion is $e^{-i\pi\sigma_x/4} e^{-i\pi\sigma_y/2} e^{-i\pi\sigma_x/4}$. One can find the following identity:

$$e^{-i\pi\sigma_x/4} e^{-i\pi\sigma_y/2} e^{-i\pi\sigma_x/4} = e^{-i\pi\sigma_y/2}. \tag{1}$$

This is the combination of several pulses, not a signle pulse. The total process is more accurate than the manipulation by the single pulse $e^{-i\pi\sigma_y/2}$ in experiments.[16-19] The composite pulse $\beta_a \beta'_{a'} \beta''_{a''}$ corresponds to the unitary operator

$$\mathcal{R}(\beta''; a'')\mathcal{R}(\beta'; a')\mathcal{R}(\beta; a), \tag{2}$$

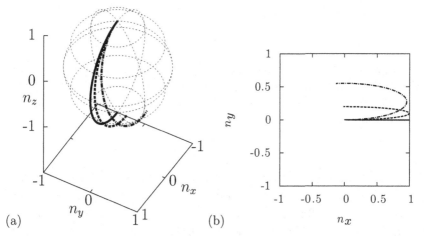

Fig. 1. Curves starting from $^t(0, 0, 1)$ (a) on the Bloch sphere and (b) on $n_x n_y$-plain during the single pulse 180_y. The solid line is the ideal case (i.e., $\Delta = 0$). The dashed line is $\Delta = 0.1$, and the dash-dotted line is $\Delta = 0.3$.

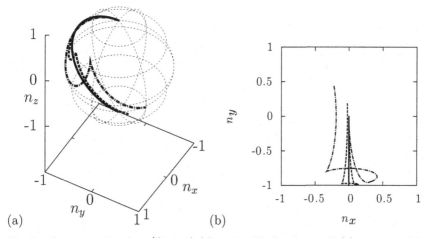

Fig. 2. Curves starting from $^t(0, 0, 1)$ (a) on the Bloch sphere and (b) on $n_x n_y$-plain during the composite pluse $90_x 180_y 90_x$. The solid line is the ideal case (i.e., $\Delta = 0$). The dashed line is $\Delta = 0.1$, and the dash-dotted line is $\Delta = 0.3$.

where $\mathcal{R}(\beta; a) = e^{-i\beta\sigma_a/2}$. The above compensated population inversion is written by $90_x 180_y 90_x$. In addition, the pulse $\bar{\beta}_a$ means the unitary operator $\mathcal{R}^\dagger(\beta; a) = \mathcal{R}(-\beta; a)$. The angle β in $\mathcal{R}(\beta; a)$ should be in the unit of radian, but in the pulse β_a we write the corresponding value in the unit of degree. We observe the robustness of $90_x 180_y 90_x$ against an undesired error. Let us consider the Hamiltonian

$$H_{n1}(t) = \frac{1}{2}\pi(\sigma_y + \Delta\sigma_z) \quad (0 \le t \le 1). \tag{3}$$

The paths of the Bloch vector starting from $^t(0,0,1)$ associated with $H_{n1}(t)$ for $\Delta = 0$, 0.1, and 0.3 are shown in Fig. 1. The presence of σ_z significantly affects the final Bloch vector at $t = 1$. Next, we consider the time evolution by the Hamiltonian

$$H_{n2}(t) = H_0(t) + \frac{1}{2}\pi\Delta\sigma_z \quad (0 \le t \le 1), \tag{4}$$

where the first and the second terms describe the Hamiltonian corresponding to $90_x 180_y 90_x$ and the error, respectively. From Fig. 2, we find that the error has less influence on the final position on the Bloch vector for the small value of Δ. They implies $90_x 180_y 90_x$ is a more robust operation against an undesired σ_z term than 180_y. This term appears when the frequency of a rf-pulse differs form the Larmor frequency of the nucleus in the field of NMR.

2.2. Construction of composite pulses

Various composite pulses have been constructed, depending on the errors which one wants to compensate. First, we explain the geometrical approach.[16–19] We visualize the trajectory of the Bloch vector, like Figs. 1 and 2. Choosing an initial point (e.g., north pole) and a target point (e.g., south pole), one can draws the trajectory on the Bloch sphere. Thus, the optimal pulse sequences such that the Bloch vector at the final time is very close to the target point are constructed.

A systematic construction is the propagator compensation through the Magunus expansion.[16–19] One defect of the geometrical approach is that the effectiveness of the composite pulses depends on the initial states. In the propagator compensation, one searches an approximate unitary operator for a target unitary operator. Let us write the time evolution operator (i.e., propagator) in the presence of errors as W. By the Magunus expansion,[22] an approximate expression for W can be obtained: $W \approx W_{\text{id}}W_{\text{err}}$. The propagator W_{id} is the time evolution operator without the errors (i.e., the ideal evolution). The propagator W_{err} represents the effect of the errors. Accordingly, a composite pulse can be obtained if one vanishes or decreases the contribution from W_{err}.

It is well known that reliable operations are made by using a simple composite pulse cyclically. These pulse sequences are called supercycles.[16–19]

Let us write a composite pulse as R. Then, we define the cycles, C and P as $C = RR\bar{R}\bar{R}$ and $P = \bar{R}RR\bar{R}$, respectively. The pulse sequence $\bar{R}$ is derived from R by inverting the rf-phases in every rf pulse (i.e., phase inversion). If $R = 90_x\overline{180}_x270_x$, for example, then $\bar{R} = \overline{90}_x180_x\overline{270}_x$. The pulse sequences, C, CP and $CP\bar{C}\bar{P}$ are examples of supercycles. When $R = 90_x180_y90_x$, the corresponding cyclic sequences are the MLEV sequences.[16–19] In particular, C and $CP\bar{C}\bar{P}$ are called MLEV-4 and MLEV-16, respectively. On the other hand, when $R = 90_x\overline{180}_x270_x$, the corresponding sequences are the WALTZ sequences.[16–19] In this case, the cycle C is called WALTZ-4, for example.

2.3. Application to quantum information processing

Finally, we remark on the application of the above traditional method to quantum information processing. Recently, the composite pulse was applied in the context of NMR quantum computation. Cummins and the collaborators[23,24] showed the composite pulses are useful to reduce the systematic errors in quantum gates. Their approach is based on the propagator compensation. Several researchers[25,26] discussed the effectiveness of the composite pulse approach to eliminate leakage errors, which lead to the non-zero transition from the encoding subspace of qubits to the other subspace (e.g., a higher energy level). The error elimination by the composite pulses may be a universal method. Namely, their effectiveness doesn't depend on the physical system. In fact, the application to the atomic qubits and the trapped ions was reported.[27,28] We hope that the reinvestigation of the composite pulses is important to establish a reliable control scheme of quantum systems and realize a genuine quantum computer.

3. Geometric phases by Aharonov and Anandan

3.1. Aharonov-Anandon phase

We review the geometric phase which was introduced by Aharonov and Anandon[20] (Aharanov-Anandan phase). Let us consider a state vector $|\psi\rangle \in \mathcal{H}$ in a quantum system, where $\mathcal{H}$ is the corresponding Hilbert space. We assume the dimension of $\mathcal{H}$ is finite. The Hamiltonian at time t is given by $H(t)$ ($0 \leq t \leq T$). Hereafter, we use the unit system in which the Planck constant $\hbar$ is equal to one. Suppose that the vector $|\psi\rangle$ is transformed into $|\psi(T)\rangle = e^{i\gamma}|\psi\rangle$ ($\gamma \in \mathbb{R}$) by the Schrödinger equation

$$i\frac{d}{dt}|\psi(t)\rangle = H(t)|\psi(t)\rangle, \tag{5}$$

where $|\psi(0)\rangle = |\psi\rangle$. The final state $|\psi(T)\rangle$ is identified with $|\psi\rangle$, except for the overall phase factor $e^{i\gamma}$. Two vectors $|u\rangle, |v\rangle \in \mathcal{H}$ are physically equivalent when there is a complex number $c \in \mathbb{C}$ such that $|c| = 1$ and $|u\rangle = c|v\rangle$. Thus, we can introduce the equivalent relation $\sim$. If $|u\rangle \sim |v\rangle$, then there exists $c \in \mathbb{C}$ such that $|c| = 1$ and $|u\rangle = c|v\rangle$. Let us introduce a surjective map $\mathcal{P}$ from $\mathcal{H}$ to $\mathcal{P}(\mathcal{H}) \equiv \mathcal{H}/\sim$. The Hilbert space $\mathcal{P}(\mathcal{H})$ is called a projective Hilbert space of $\mathcal{H}$.[20,43] The temporal behavior of the vector $|\psi\rangle$ is regarded as a cyclic evolution in $\mathcal{P}(\mathcal{H})$. Namely, for a curve $C = \{t \in [0, T] \mapsto |\psi(t)\rangle \in \mathcal{H}\} \subset \mathcal{H}$, we find the closed loop $\hat{C} \equiv \mathcal{P}(C)$ in $\mathcal{P}(\mathcal{H})$. Let us define a new vector $|\psi(t)\rangle\rangle \in \mathcal{H}$ such that

$$|\psi(T)\rangle\rangle = |\psi(0)\rangle\rangle, \tag{6}$$

and

$$|\psi(t)\rangle\rangle = e^{-iF(t)}|\psi(t)\rangle, \tag{7}$$

where $F(t)$ is a real smooth function with respect to $t \in [0, T]$. From Eqs. (5) and (7), we obtain $-\dot{F}(t)\, e^{iF(t)}|\psi(t)\rangle\rangle + ie^{iF(t)}\frac{d}{dt}|\psi(t)\rangle\rangle = H(t)e^{iF(t)}|\psi(t)\rangle\rangle$. Accordingly, we find

$$-F(T) + F(0) + i\int_0^T \langle\langle\psi(t)|\frac{d}{dt}|\psi(t)\rangle\rangle\, dt = \int_0^T \langle\psi(t)|H(t)|\psi(t)\rangle\, dt. \tag{8}$$

From the requirement (6), we have

$$|\psi(T)\rangle = e^{iF(T)}|\psi(T)\rangle\rangle = e^{iF(t)}|\psi(0)\rangle\rangle = e^{i(F(T)-F(0))}|\psi\rangle, \tag{9}$$

while we have supposed $|\psi(T)\rangle = e^{i\gamma}|\psi\rangle$. Therefore, we find that $\gamma = F(T) - F(0)$, up to $2\pi\kappa$ ($\kappa \in \mathbb{Z}$). Consequently, we obtain

$$\gamma = \gamma_d + \gamma_g, \tag{10}$$

where

$$\gamma_d = -\int_0^T \langle\psi(t)|H(t)|\psi(t)\rangle\, dt, \tag{11}$$

$$\gamma_g = i\int_0^T \langle\langle\psi(t)|\frac{d}{dt}|\psi(t)\rangle\rangle\, dt. \tag{12}$$

The phase γ_d represents the dynamical phase, while the phase γ_g is the geometric phase introduced by Aharanov and Anandan[20] and called an Aharanov-Anandan (AA) phase. The AA phase is the generalization of a Berry phase,[29,30] which arises under an adiabatic and periodic control process, to a non-adiabatic control process.

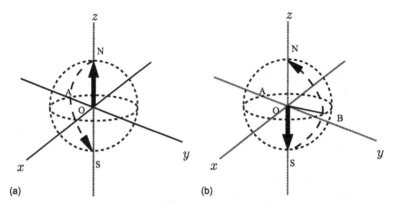

Fig. 3. Temporal behavior of $n = {}^t(0, 0, 1)$ for the Hamiltonian (14). The points A and B indicate ${}^t(0, -1, 0)$ and ${}^t(-\sin\phi, \cos\phi, 0)$, respectively. The point N (S) means the north (south) pole on the Bloch sphere. The closed loop given by the motion of the Bloch vector is NASBN. Note that the angle $\angle AOB$ is $\pi - \phi$. (a) $0 \le t \le \pi/\omega_1$ (b) $\pi/\omega_1 \le t \le T$.

The AA phase, which is different from the dynamical phase γ_d, a geometrical object and depends on only a geometrical property of the closed loop $\hat{C}$. Actually, a connection in a principle bundle plays an essential role for its description.[31] This point was discussed in Ref. 32 by using the differential geometry. In this lecture note, we will observe this fact by discussing the following two points.

First of all, the value of γ_g for a given closed loop $\hat{C}$ is independent of the choice of the parameterization of $\hat{C}$. Let us consider the new parameterization $\tau = \tau(t)$ in $\hat{C}$, where $\tau(t)$ is invertible and $|\psi[\tau(T)]\rangle = |\psi[\tau(0)]\rangle$. Obviously

$$i \int_0^T \langle\langle \psi(t)| \frac{d}{dt} |\psi(t)\rangle\rangle \, dt = i \int_{\tau(0)}^{\tau(T)} \langle\langle \psi(\tau)| \frac{d}{d\tau} |\psi(\tau)\rangle\rangle \, d\tau. \tag{13}$$

Secondly, we investigate the relation of the value of γ_g to the geometrical property of $\hat{C}$. Let us consider the temporal behavior of a two-level quantum system described by the following Hamiltonian:

$$H(t) = \begin{cases} \pi\sigma_x/2 & (0 \le t \le 1/2) \\ \pi(\sigma_x \cos\phi + \sigma_y \sin\phi)/2 & (1/2 \le t \le 1). \end{cases} \tag{14}$$

The matrix σ_a represents the a-component of the Pauli matrix ($a = x, y, z$). We write the eigenvectors of σ_z as $|0\rangle$ and $|1\rangle$, where $\sigma_z|0\rangle = |0\rangle$ and $\sigma_z|1\rangle = -|1\rangle$. In this example, the Hilbert space $\mathcal{H}$ is $\mathbb{C}^2$. Let us introduce the Bloch vector $\boldsymbol{n}(t)$ corresponding to $|\psi(t)\rangle \in \mathbb{C}^2$ as

$$|\psi(t)\rangle\langle\psi(t)| = \frac{1}{2}\mathbb{1} + \frac{1}{2}\boldsymbol{n}(t) \cdot \boldsymbol{\sigma}, \tag{15}$$

where $\mathbb{1}$ means the 2×2 identity matrix, $\boldsymbol{\sigma} = {}^t(\sigma_x, \sigma_y, \sigma_z)$, and $\boldsymbol{n}(t) = \langle\psi(t)|\boldsymbol{\sigma}|\psi(t)\rangle$. When the system is pure, the length of the corresponding Bloch vector is unity. The temporal behavior of $|\psi(t)\rangle$ can be drawn as a trajectory on a unit sphere in $\mathbb{R}^3$. Namely, $\boldsymbol{n}(t) \in S^2 = \{\boldsymbol{n} \in \mathbb{R}^3; |\boldsymbol{n}| = 1\}$. The Bloch vector for $c|\psi\rangle$ is the same as for $|\psi\rangle$, where c is a complex number with $|c| = 1$. Thus, we find the projective Hilbert space in this example is S^2. We assume that the initial Bloch vector $\boldsymbol{n}(0)$ is ${}^t(0, 0, 1)$ (i.e., $|\psi(0)\rangle = |0\rangle$). Furthermore, we assume that $0 \leq \phi \leq \pi$. We readily check that $\boldsymbol{n}(1) = \boldsymbol{n}(0)$. The state vector at $t = 1$ is

$$|\psi(1)\rangle = e^{i(\pi-\phi)}|0\rangle. \tag{16}$$

Next, let us calculate the dynamical phase accompanying with this time evolution. Generally, the Hamiltonian in a two-level system can be expressed by

$$H(t) = \frac{1}{2}\boldsymbol{h}(t) \cdot \boldsymbol{\sigma}, \tag{17}$$

where $\boldsymbol{h}(t) \in \mathbb{R}^3$. Remark that the trace of the Hamiltonian can be zero without losing generality. The temporal behavior of $\boldsymbol{n}(t)$ is described by the Bloch equation:

$$\frac{d}{dt}\boldsymbol{n}(t) = \boldsymbol{h}(t) \times \boldsymbol{n}(t). \tag{18}$$

If the Hamiltonian (17) is independent of time (i.e., $\boldsymbol{h}(t) = \boldsymbol{h}$), we find

$$\frac{d}{dt}\boldsymbol{h} \cdot \boldsymbol{n}(t) = \boldsymbol{h} \cdot (\boldsymbol{h} \times \boldsymbol{n}(t)) = 0, \tag{19}$$

from Eq. (18). We calculate the expectation value of $H(t)$ to obtain the dynamical phase from Eq. (11): $\langle\psi(t)|H(t)|\psi(t)\rangle = \frac{1}{2}\boldsymbol{h}(t) \cdot \boldsymbol{n}(t)$. The Hamiltonian (14) is independent of time in either $[0, 1/2]$ or $[1/2, 1]$. The dynamical phase in this example is

$$\begin{aligned}
\gamma_{\mathrm{d}} &= -\frac{1}{2}{}^t(\pi\cos\phi, \pi\sin\phi, 0) \cdot \boldsymbol{n}(1/2) - \frac{1}{2}{}^t(\pi, 0, 0) \cdot \boldsymbol{n}(0) \\
&= -\frac{1}{2}{}^t(\pi\cos\phi, \pi\sin\phi, 0) \cdot {}^t(0, 0, -1) - \frac{1}{2}{}^t(\pi, 0, 0) \cdot {}^t(0, 0, 1) \\
&= 0.
\end{aligned}$$

Therefore, the AA phase acquired is $\pi - \phi$. Remark that the solid angle of the sphere skullcap given by the closed loop on S^2 is $2(\pi - \phi)$. Thus, we find the AA phase is proportional to the solid angle [Fig. 3].

The previous example implies the condition for the dynamical phase to be zero. Namely, if the curve C satisfies

$$\langle \psi(t)|H(t)|\psi(t)\rangle = 0 \quad (0 \le t \le T), \tag{20}$$

$\gamma_d = 0$ (i.e., $\gamma = \gamma_g$). Or, using Eq. (5) and the fact that $\langle \psi(t)|\psi(t)\rangle$ is constant, the above condition can be rewritten as

$$\text{Im}\,\langle \psi(t)|\frac{d}{dt}|\psi(t)\rangle = 0 \quad (0 \le t \le T). \tag{21}$$

3.2. Further generalization of geometric phases

It is necessary for defining the AA phase to impose that a quantum system should undergo a cyclic evolution. However, it is not possible to completely implement the control processes by which the cyclic evolution is accomplished. In addition, a quantum system is not always an isolated system; it is suffered from decoherence. Thus, it is important to define a geometric phase when the system undergoes a non-cyclic or non-unitary time evolution. Such a topic must be related to the robustness of the geometric phases and the quamtum gates based on them. We will write several useful references for further development on this issue below.

The geometric phase in the case of non-cyclic time evolution was researched by Samuel and Bhandari.[33] Their consideration is based on the Pancharatnam phase difference between two non-orthogonal vectors.[34,35] Their formalism was applied to the research about the effect of fluctuations on the Berry phase of a two-level system.[10] In addition, the application to non-adiabatic non-cyclic time evolution of a two-level system was also investigated in Ref. 36.

Holonomy for arbitrary density matrices, which is generally mixed, was proposed by Uhlmann.[37,38] The basic idea is to consider the purification of a given density matrix.[21] Let us write the set of all density matrices on the Hilbert space $\mathcal{H}$ as $\mathcal{S}$. To achieve purification, we have to consider an extended Hilbert space $\mathcal{K} = \mathcal{H} \otimes \mathcal{H}'$, where $\mathcal{H}'$ is the Hilbert space for an ancillary quantum system. If we choose the dual Hilbert space $\mathcal{H}^*$ for $\mathcal{H}$ as $\mathcal{H}'$, we can find $\mathcal{K}$ is the Hilbert space of the Hilbert-Schmidt operators.[38] Note that the purification of a density matrix is not uniquely defined. Namely, for a given density matrix $\rho \in \mathcal{S}$, two (or more than two) possible solutions exist. It implies that we can define a surjective map $\mathcal{P}$:

$\mathcal{K} \to \mathcal{S}$. Thus, we can naturally introduce a bundle structure[31] and define holonomy for the general density matrices. On the other hand, based on the Pancharatnam connection,[34,35] the theory[39,40] and the experiments[41] of geometric phases for density matrices have been reported. Furthermore, the connection between non-unitary time evolution (i.e., a complete positive map) and geometric phases of mixed states was discussed in Ref. 42. Their formalism will be important to examine the stability of geometric phases against decoherence.[14]

Finally, we remark on the other important related topics. The study for the geometric phases in the general situation is deeply connected with the fundamental investigation of quantum mechanics, the geometrical structure of quantum states.[43,44] Uhlmann's work may give us a very beautiful mathematical scheme to understand the geometric property of $\mathcal{S}$.

4. Non-adiabatic Geometric Quantum Gates

We explain how to construuct a one-qubit GQG with an AA phase (i.e., a non-adiabatic GQG).[4] Let us write the Bloch vector at t $(0 \le t \le 1)$ as $\boldsymbol{n}(t)(\in \mathbb{R}^3)$. We denote a state vector given $\boldsymbol{n}(t)$ as $|\boldsymbol{n}(t)\rangle(\in \mathbb{C}^2)$. Thus

$$\boldsymbol{n}(t) = \langle \boldsymbol{n}(t)|\boldsymbol{\sigma}|\boldsymbol{n}(t)\rangle. \tag{22}$$

The Hamiltonian of the system is written as $H(t)$ $(0 \le t \le 1)$. We put $T = 1$. Suppose that there exist two normalized Bloch vectors $\boldsymbol{n}_\pm$ which satisfy two conditions,

(a) $\boldsymbol{n}_+ \cdot \boldsymbol{n}_- = -1$ (i.e., $\langle \boldsymbol{n}_+|\boldsymbol{n}_-\rangle = 0$).
(b) there exist $\gamma_\pm \in \mathbb{R}$ such that $|\boldsymbol{n}_\pm(1)\rangle = e^{i\gamma_\pm}|\boldsymbol{n}_\pm\rangle$ (i.e., $\boldsymbol{n}_\pm(1) = \boldsymbol{n}_\pm$).

An arbitrary quantum state $|\boldsymbol{n}\rangle$ for a one-qubit can be expressed by

$$|\boldsymbol{n}\rangle = a_+|\boldsymbol{n}_+\rangle + a_-|\boldsymbol{n}_-\rangle, \tag{23}$$

where $a_\pm = \langle \boldsymbol{n}_\pm|\boldsymbol{n}\rangle$ since $\langle \boldsymbol{n}_+|\boldsymbol{n}_-\rangle = 0$. We call $\boldsymbol{n}_\pm$ the basis Bloch vectors corresponding to $H(t)$. The initial state $|\boldsymbol{n}\rangle$ is transformed into the final state

$$|\boldsymbol{n}(1)\rangle = a_+ e^{i\gamma_+}|\boldsymbol{n}_+\rangle + a_- e^{i\gamma_-}|\boldsymbol{n}_-\rangle. \tag{24}$$

Therefore, the time evolution operator U at $t = 1$ is

$$U = \mathcal{T} \exp\left[-i \int_0^1 H(t)dt\right] = e^{i\gamma_+}|\boldsymbol{n}_+\rangle\langle \boldsymbol{n}_+| + e^{i\gamma_-}|\boldsymbol{n}_-\rangle\langle \boldsymbol{n}_-|, \tag{25}$$

where the symbol $\mathcal{T}$ means the time ordering operator. Obviously, $|\boldsymbol{n}(1)\rangle = U|\boldsymbol{n}\rangle$. Equation (25) represents a one-qubit quantum gate constructed by the geometric phases $\gamma_\pm$, if the dynamical components of $\gamma_\pm$ are vanishing.

Taking account of a NMR system, let us focus on the following Hamiltonian for a one-qubit system:

$$H(t) = \frac{1}{2}\omega(t)\,\boldsymbol{m}(t)\cdot\boldsymbol{\sigma} \qquad (0 \le t \le 1), \tag{26}$$

where

$$\boldsymbol{m}(t) = \begin{pmatrix} \sin\chi(t)\cos\phi(t) \\ \sin\chi(t)\sin\phi(t) \\ \cos\chi(t) \end{pmatrix}. \tag{27}$$

Here, $\omega(t)$ and $\boldsymbol{m}(t)$ are the amplitude of and the unit vector parallel to a magnetic field, respectively. The phase $\phi(t)$ corresponds to the rf-phase. We have already derived the condition for the dynamical phase to be vanishing in Sec. 3. From Eqs. (20) and (26), we find the condition is given by

$$\boldsymbol{n}(t)\cdot\boldsymbol{m}(t) = 0 \quad (0 \le t \le 1). \tag{28}$$

We note tha double-loop method for removing a dymanical phase during gate operation is commonly employed.[3,7] Namely, The geometric phases in the first and second steps are additive, while the dynamical phases are canceled with each other.

Let us rewirte Eq. (28) as a practical form. In the typical experiments of NMR, pulse sequences are employed. Therefore, we can assume the time-dependence of $\omega(t)$, $\chi(t)$, and $\phi(t)$ should be given as

$$\omega(t) = \sum_{i=1}^{N} \omega_i\, w(t; t_i, t_{i-1}) \quad (\omega_i \in \mathbb{R}), \tag{29}$$

$$\chi(t) = \sum_{i=1}^{N} \chi_i\, w(t; t_i, t_{i-1}) \quad (\chi_i \in \mathbb{R}), \tag{30}$$

$$\phi(t) = \sum_{i=1}^{N} \phi_i\, w(t; t_i, t_{i-1}) \quad (\phi_i \in \mathbb{R}), \tag{31}$$

respectively. Note that $t_N = 1$ and $t_0 = 0$. In addition, if $i > j$, then $t_i > t_j$. The definition of the function $w(t; t_i, t_{i-1})$ is given as

$$w(t; t_i, t_{i-1}) = \begin{cases} 1 & (t_{i-1} \le t \le t_i) \\ 0 & (\text{otherwise}) \end{cases}. \tag{32}$$

From Eq. (11), the dynamical phase for $|n_\pm\rangle$ during the total process is

$$\gamma_{\pm d} = -\sum_{i=1}^{N} \theta_i \, m_i \cdot n_\pm(t_{i-1}), \tag{33}$$

where

$$\theta_i = \omega_i(t_i - t_{i-1}), \tag{34}$$

$$m_i = \begin{pmatrix} \sin\chi_i \cos\phi_i \\ \sin\chi_i \sin\phi_i \\ \cos\chi_i \end{pmatrix}. \tag{35}$$

We readily find that

$$\frac{d}{dt} m_i \cdot n_\pm(t) = 0 \quad (t_{i-1} \le t \le t_i). \tag{36}$$

Accordingly, we obtain a necesarry condition for $\gamma_{+d} = 0$ as follows:

$$m_i \cdot n_+(t_{i-1}) = 0 \quad (i = 1, 2, \ldots, N). \tag{37}$$

Equation (37) implies $m_i \cdot n_-(t_{i-1}) = 0$ for the arbitrary i, because $n_- = -n_+$. If the Bloch vector is perpendicular to the magnetic field in the rotating frame, no dynamical phase exists.[45] This criterion is simple and practical. Actually, it has been used in many experiments on geometric phases.[4–6,45] In the current model, the left hand side in Eq. (25) is rewritten in term of pulse sequences:

$$U = V(t_N, t_{N-1}) \ldots V(t_1, t_0), \tag{38}$$

where $V(t_i, t_{i-1}) = e^{-i\theta_i m_i \cdot \sigma/2}$. Equation (38) implies an experimental process to implement geometric quantum gates given by the right hand side of Eq. (25).

5. Reinterpretation of Composite Pulses

We show that several composite pulses are interpreted as geometrical quantum gates.[15] We discuss two types of composite pulses: $90_x 180_y 90_x$ and $90_x \overline{180}_x 270_x$. The former is the elementary unit of MLEV sequences and the latter is the elementary one of WALTZ sequences.[16–19] What is an important thing is that we reconstruct such pulse sequences from the viewpoint of quantum gates. Namely, our discussion does not depend on the specific choice of an initial quantum state.

5.1. $90_x 180_y 90_x$

We discuss the composite pulses $90_x 180_y 90_x$. A series of pulses $90_x 180_y 90_x$[16,19] is widely employed in the field of NMR. Its corresponding unitary operator is $e^{-i\pi\sigma_x/4} e^{-i\pi\sigma_y/2} e^{-i\pi\sigma_x/4}$. The above series of pulses, employed for wide band decoupling,[19] is equivalent to 180_y. This composite pulse is generated by the Hamiltonian

$$H(t) = \pi \boldsymbol{m}(t) \cdot \boldsymbol{\sigma}, \tag{39}$$

where

$$\boldsymbol{m}(t) = \begin{cases} {}^t(1,\,0,\,0) & (t_0 \le t \le t_1) \\ {}^t(0,\,1,\,0) & (t_1 \le t \le t_2) \\ {}^t(1,\,0,\,0) & (t_2 \le t \le t_3) \end{cases}, \tag{40}$$

$t_0 = 0$, $t_1 = 1/4$, $t_2 = 3/4$, and $t_3 = 1$. Compared to Eqs. (29)-(31), we find that

$$\omega_1 = \omega_2 = \omega_3 = 2\pi, \tag{41}$$

$$\chi_1 = \chi_2 = \chi_3 = \frac{\pi}{2}, \tag{42}$$

$$\phi_1 = \phi_3 = 0, \quad \phi_2 = \frac{\pi}{2}. \tag{43}$$

From Eq. (38), we can readily find that the time evolution generated by Eq. (39) is the aimed composite pulse. Let us put $\boldsymbol{n}_\pm = \pm {}^t(0,\,1,\,0)$. Obviously, $\boldsymbol{n}_+ \cdot \boldsymbol{n}_- = -1$. The vector $|n_\pm\rangle$ corresponding to $\boldsymbol{n}_\pm$ is given by $|n_\pm\rangle = (|0\rangle \pm i|1\rangle)/\sqrt{2}$, up to the overall phase. The explicit expression for $\boldsymbol{n}_\pm(t)$ is given as

$$\boldsymbol{n}_\pm(t) = \pm \begin{pmatrix} \sin\theta(t) \sin\phi(t) \\ -\sin\theta(t)\cos\phi(t) \\ \cos\theta(t) \end{pmatrix}, \tag{44}$$

where

$$\theta(t) = 2\pi t - \frac{\pi}{2}, \tag{45}$$

$$\phi(t) = \begin{cases} 0 & (t_0 \le t \le t_1) \\ \pi/2 & (t_1 \le t \le t_2) \\ 0 & (t_2 \le t \le t_3) \end{cases}. \tag{46}$$

Note that $\boldsymbol{n}_-(t) = -\boldsymbol{n}_+(t)$. We can readily check $\boldsymbol{m}_i \cdot \boldsymbol{n}_\pm(t_{i-1}) = 0$ ($i = 2,\,3$) and $\boldsymbol{n}_\pm(1) = \boldsymbol{n}_\pm$ [Table 1]. Next, we evaluate the value of $\gamma_\pm$. We find

138

Table 1. Parameters and temporal behavior of the basis Bloch vector during $90_x180_y90_x$

i	θ_i	ϕ_i	$\boldsymbol{m}_i$	$\boldsymbol{n}_\pm(t_i)$
0	-	-	-	$\pm(0, 1, 0)$
1	$\pi/2$	0	$(1, 0, 0)$	$\pm(0, 0, 1)$
2	π	$\pi/2$	$(0, 1, 0)$	$\pm(0, 0, -1)$
3	$\pi/2$	0	$(1, 0, 0)$	$\pm(0, 1, 0)$

that

$$
\begin{aligned}
|\boldsymbol{n}_+(1)\rangle &= e^{-i\pi\sigma_x/4}e^{-i\pi\sigma_y/2}e^{-i\pi\sigma_x/4}|\boldsymbol{n}_+\rangle \\
&= e^{-i\pi\sigma_x/4}e^{-i\pi\sigma_y/2}|0\rangle \\
&= e^{-i\pi\sigma_x/4}(-i)i|1\rangle \\
&= (-i)i\frac{1}{\sqrt{2}}(|1\rangle - i|0\rangle) = (-i)|\boldsymbol{n}_+\rangle,
\end{aligned}
$$

and

$$
\begin{aligned}
|\boldsymbol{n}_-(1)\rangle &= e^{-i\pi\sigma_x/4}e^{-i\pi\sigma_y/2}e^{-i\pi\sigma_x/4}|\boldsymbol{n}_-\rangle \\
&= e^{-i\pi\sigma_x/4}e^{-i\pi\sigma_y/2}(-i)|1\rangle \\
&= e^{-i\pi\sigma_x/4}(-i)^2(-i)|0\rangle \\
&= (-i)^3\frac{1}{\sqrt{2}}(|0\rangle - i|1\rangle) = i|\boldsymbol{n}_-\rangle.
\end{aligned}
$$

See, Table 1, as well. Thus, we have

$$|\boldsymbol{n}_\pm(1)\rangle = e^{\mp i\pi/2}|\boldsymbol{n}_\pm\rangle, \qquad (47)$$

and $\gamma_\pm = \mp\pi/2$. The resultant non-adiabatic geometric quantum gate is $U(1) = e^{-i\pi\sigma_y/2}$ from Eq. (25). Therefore, the composite pulse $90_x180_y90_x$ is just a quantum gate constructed by pure geometric phases. Other possibility also exists. We can find the composite pulse $90_x\overline{180}_y90_x$ is also a non-adiabatic geometric quantum gate. Moreover, we can find that geometric phases have been used in the cyclic composite pulses whose elementary unit is $90_x180_y90_x$ (e.g., MLEV-4[16–19]).

Finally, we would like to explain the evaluation of the geometric phase based on Eq. (12). In the above argument, we have obtained it by following the temporal evolution of the state vector $|\boldsymbol{n}_\pm\rangle$ during $90_x180_y90_x$. On the other hand, using Eq. (12), we can derive it as a solid angle of the sphere skullcap given by the closed curve $\boldsymbol{n}_\pm(t)$ on the Bloch sphere. For this purpose, the vector which satisfies Eq. (6) is necessary. Let us introduce

$|n_\pm(t)\rangle\rangle$ as

$$|n_+(t)\rangle\rangle = \cos\frac{\Theta(t)}{2}|e_3+\rangle + \sin\frac{\Theta(t)}{2}e^{i\Phi(t)}|e_3-\rangle, \tag{48}$$

$$|n_-(t)\rangle\rangle = -\sin\frac{\Theta(t)}{2}|e_3+\rangle + \cos\frac{\Theta(t)}{2}e^{i\Phi(t)}|e_3-\rangle, \tag{49}$$

where

$$|e_3+\rangle = \cos\frac{\pi}{4}|0\rangle + e^{i\pi/4}\sin\frac{\pi}{4}|1\rangle, \tag{50}$$

$$|e_3-\rangle = -\sin\frac{\pi}{4}|0\rangle + e^{i\pi/4}\cos\frac{\pi}{4}|1\rangle, \tag{51}$$

$\cos\Theta(t) = n_+(t)\cdot e_3,, |\sin\Theta(t)|\cos\Phi(t) = n_+(t)\cdot e_1$, and $|\sin\Theta(t)|\sin\Phi(t) = n_+(t)\cdot e_2$. The real vectors e_1, e_2, and e_3 are defined as, respectively,

$$e_1 = {}^t(0,0,-1), \quad e_2 = \frac{1}{\sqrt{2}}{}^t(-1,1,0), \quad e_3 = \frac{1}{\sqrt{2}}{}^t(1,1,0).$$

We readily find that $e_3 = \langle e_3+|\boldsymbol{\sigma}|e_3+\rangle$. Remark that the domain of the polar angle Θ is $0 \le \Theta \le \pi$. Thus, we find that $\sin\Theta \ge 0$. We show that $|n_\pm(1)\rangle\rangle = |n_\pm(0)\rangle\rangle$. Recall that $n_+(1) = n_+(0)$. Thus $n_+(1)\cdot e_3 = n_+(0)\cdot e_3$. It means $\cos\Theta(1) = \cos\Theta(0)$. From $\sin\Theta(t) \ge 0$, we find $\Theta(1) = \Theta(0)$. Similarly, $n_+(1)\cdot e_1 = n_+(0)\cdot e_1$ and $n_+(1)\cdot e_2 = n_+(0)\cdot e_2$. It means that $\cos\Phi(1) = \cos\Phi(0)$ and $\sin\Phi(1) = \sin\Phi(0)$ because $\sin\Theta(1) = \sin\Theta(0)$. Therefore $\Phi(1) = \Phi(0)$. Now, let us show the explicit expression for $\gamma_{\pm g}$. From Eq. (12)

$$\begin{aligned}
\gamma_{g\pm} = -\frac{1}{2}\Bigg[&\int_0^{1/4} \frac{1}{1\mp\sin\theta(t)\cos(\phi(t)+\pi/4)}\cos(\phi(t)-\pi/4)\frac{d\theta}{dt}dt \\
&+ \int_{1/4}^{3/4} \frac{1}{1\mp\sin\theta(t)\cos(\phi(t)+\pi/4)}\cos(\phi(t)-\pi/4)\frac{d\theta}{dt}dt \\
&+ \int_{3/4}^1 \frac{1}{1\mp\sin\theta(t)\cos(\phi(t)+\pi/4)}\cos(\phi(t)-\pi/4)\frac{d\theta}{dt}dt \Bigg] \\
= -\frac{1}{2}\Bigg[&\int_{-\pi/2}^0 \frac{1}{1\mp\frac{1}{\sqrt{2}}\sin\theta}\frac{1}{\sqrt{2}}d\theta + \int_0^\pi \frac{1}{1\mp\frac{-1}{\sqrt{2}}\sin\theta}\frac{1}{\sqrt{2}}d\theta \\
&+ \int_\pi^{3\pi/2} \frac{1}{1\mp\frac{1}{\sqrt{2}}\sin\theta}\frac{1}{\sqrt{2}}d\theta \Bigg].
\end{aligned}$$

140

Then

$$\gamma_{+g} = -\frac{1}{2}\left(\frac{\pi}{4} + \frac{\pi}{2} + \frac{\pi}{4}\right) = -\frac{\pi}{2}, \tag{52}$$

$$\gamma_{-g} = -\frac{1}{2}\left(\frac{3\pi}{4} + \frac{3\pi}{2} + \frac{3\pi}{4}\right) = -\frac{3\pi}{2} = \frac{\pi}{2} - 2\pi \equiv \frac{\pi}{2}. \tag{53}$$

Theses results are completely equal to the previous ones.

5.2. $90_x\overline{180}_x270_x$

Let us consdier another type of composite pulses which is given by $90_x\overline{180}_x270_x$ (i.e., $\mathcal{R}\left(\frac{3\pi}{2}; x\right)\mathcal{R}\left(-\pi; x\right)\mathcal{R}\left(\frac{\pi}{2}; x\right)$). Such a pulse sequence is often written by $1\bar{2}3$. Obviously, this pulse sequence is mathematically equivalent to 180_x. This composite pulse is generated by the Hamiltonian

$$H(t) = \frac{3}{2}\pi \boldsymbol{m}(t) \cdot \boldsymbol{\sigma}, \tag{54}$$

where

$$\boldsymbol{m}(t) = \begin{cases} {}^t(1, 0, 0) & (t_0 \leq t \leq t_1) \\ {}^t(-1, 0, 0) & (t_1 \leq t \leq t_2) \\ {}^t(1, 0, 0) & (t_2 \leq t \leq t_3) \end{cases}, \tag{55}$$

where $t_0 = 0$, $t_1 = 1/6$, $t_2 = 1/2$, and $t_3 = 1$. Compared to Eqs. (29)-(31), we find that

$$\omega_1 = \omega_2 = \omega_3 = 3\pi, \tag{56}$$

$$\chi_1 = \chi_2 = \chi_3 = \frac{\pi}{2}, \tag{57}$$

$$\phi_1 = \phi_3 = 0, \quad \phi_2 = \pi. \tag{58}$$

The pulse sequence $1\bar{2}3$ does not correspond to a GQG. Actually, the temporal behavior is not cyclic evolution. It should be noted that we have to define $\boldsymbol{n}_\pm$ as $\pm^t(0, \sin\varphi, \cos\varphi)$. The degrees of freedom to choose φ remains. However, the latter discussion does not depend on the value of φ. We focus on the motion of $\boldsymbol{n}_+$. At $t = t_1$, we find $\boldsymbol{n}_+(t_1) = {}^t(0, -\cos\varphi, \sin\varphi)$. Then, $\boldsymbol{m}_2 \cdot \boldsymbol{n}_+(t_1) = 0$. Repeating the similar calculation, we find $\boldsymbol{m}_3 \cdot \boldsymbol{n}(t_2) = 0$ and $\boldsymbol{n}_+(T) = \boldsymbol{n}_-$.

However, composite pulses made by the combination of several $1\bar{2}3$s are regarded as GQGs. Let us consider the successive application of $1\bar{2}3$, $1\bar{2}31\bar{2}3$. Mathematically, this pulse sequence is equivalent to $-\mathbb{1}$. By $1\bar{2}3$, as is shown above, the Bloch vector $\boldsymbol{n}_\pm$ is transformed into $\boldsymbol{n}_\mp$. Then, we apply the corresponding pusle sequence to $1\bar{2}3$ to $\boldsymbol{n}_\mp$, again. The temporal evolution of $\boldsymbol{n}_\pm$ during $1\bar{2}31\bar{2}3$ is shown in Table 2. Remark that $\boldsymbol{m}_i \cdot \boldsymbol{n}(t_{i-1}) = 0$

Table 2. Parameters and temporal behavior of the
basis Bloch vectors during $1\bar{2}31\bar{2}3$.

i	θ_i	ϕ_i	m_i	$n_\pm(t_i)$
0	–	–	–	$\pm(0, -\cos\varphi, \sin\varphi)$
1	$\pi/2$	0	$(1, 0, 0)$	$\pm(0, 0, 1)$
2	π	π	$(-1, 0, 0)$	$\pm(0, 0, -1)$
3	$3\pi/2$	0	$(1, 0, 0)$	$\pm(0, \cos\varphi, -\sin\varphi)$
4	$\pi/2$	0	$(1, 0, 0)$	$\pm(0, 0, -1)$
5	π	π	$(-1, 0, 0)$	$\pm(0, 0, 1)$
6	$3\pi/2$	0	$(1, 0, 0)$	$\pm(0, -\cos\varphi, \sin\varphi)$

($i = 1, 2, \ldots, 6$). It means that no dynamical phase occurs. What is the most important thing is that two orthonormal Bloch vectors come back to the original position at the final time t_6: $n_\pm(T) = n_\pm$. In addition, we find $\gamma_\pm = \pm\pi$. Therefore, the composite pulse $1\bar{2}31\bar{2}3$ is a GQG and is givn by $e^{i\pi}|n_+\rangle\langle n_+| + e^{-i\pi}|n_-\rangle\langle n_-|$, which is equivalent to $-\mathbb{1}$. Similarly, we can show $\bar{1}2\bar{3} (= \overline{90}_x 180_x \overline{270}_x)$ is not a geometrical quantum gate, but $\bar{1}2\bar{3}\bar{1}2\bar{3}$, $\bar{1}2\bar{3}1\bar{2}3$, and $1\bar{2}3\bar{1}2\bar{3}$ are regareded as the GQGs. Furthermore, we can find that geometric phases are used in the cyclic composite pulses whose elementary units are the above composite pulses (e.g., WALTZ–4[16–19]).

6. Effects of fluctuations

We discuss the effect of fluctuations in control parameters on the composite pluse $90_x 180_y 90_x$. In Appendix. B, we will show the general arguments to introduce fluctuations into pulse sequences. In order to exist a well-defined geometric phase, the path on the Bloch sphere must be closed even when the system is suffered from the fluctuation. Therefore, we select the following path to be investigated,

$$\tilde{n}_\pm(t) = \pm \begin{pmatrix} \sin(\theta(t) + f(t)) \sin(\phi(t) + g(t)) \\ -\sin(\theta(t) + f(t)) \cos(\phi(t) + g(t)) \\ \cos(\theta(t) + f(t)) \end{pmatrix}. \tag{59}$$

The functions $f(t)$ and $g(t)$ are continuous and smooth in $[0, 1]$, and satisfy

$$f(0) = g(0) = 0, \quad f(1) = g(1) = 0. \tag{60}$$

The closed paths exist in the Bloch sphere even if the system is fluctuated by $f(t)$ and $g(t)$ if the requirement (60) is satisfied. The result $\tilde{n}_\pm(1) = \tilde{n}_\pm$ implies that there exist $\tilde{\gamma}_\pm \in \mathbb{R}$ such that

$$|\tilde{n}_\pm(1)\rangle = e^{i\tilde{\gamma}_\pm}|\tilde{n}_\pm\rangle. \tag{61}$$

In the time interval $[t_{i-1}, t_i]$, $\theta(t)$ and $\phi(t)$ are smooth functions. Then, we differenciate $\tilde{n}_\pm(t)$ with repsect to t, and derive the corresponding Bloch equation. Summarizing the results for all time intervals, we find that these fluctuated curves on the Bloch sphere are generated by the Hamiltonian,

$$\tilde{H}(t) = \frac{1}{2}\tilde{\omega}(t)\,\tilde{\boldsymbol{m}}(t) \cdot \boldsymbol{\sigma} + \frac{1}{2}\frac{dg(t)}{dt}\sigma_z, \tag{62}$$

where

$$\tilde{\omega}(t) = 2\pi + \frac{df(t)}{dt}, \tag{63}$$

$$\tilde{\boldsymbol{m}}(t) = \begin{pmatrix} \cos(\phi(t) + g(t)) \\ \sin(\phi(t) + g(t)) \\ 0 \end{pmatrix}. \tag{64}$$

Note that $\tilde{H}(t) = H(t)$ when $f(t) = g(t) = 0$. The derivative of $f(t)$ and $g(t)$ means the fluctuation of the radiofrequency amplitude and the resonance off-set, respectively. The fluctuation of the radiofrequency phase is described by $g(t)$.

From Eq. (11), the dynamical phases associated with $|\tilde{\boldsymbol{n}}_\pm(t)\rangle$, $\tilde{\gamma}_{\pm d}$ are given by

$$\tilde{\gamma}_{\pm d} = \mp\frac{1}{2}\int_{t_0}^{t_3} \frac{dg(t)}{dt}\cos(\theta(t) + f(t))\,dt. \tag{65}$$

We notice that the dynamical phases are exactly vanishing if the funcitons $f(t)$ and $g(t)$ satisfy the conditions which we explain below. First, we readily find that $\tilde{\gamma}_{\pm d} = 0$ if $g(t) = 0$ for the arbitrary $t \in [0, 1]$. Next, we assume they have proper symmetry under the time translation. Namely, if

$$f(t + 1/2) = f(t), \quad g(t + 1/2) = g(t), \tag{66}$$

then we find $\tilde{\gamma}_{\pm d} = 0$ from Eq. (65). Thus, we find the fluctuation under which the dynamical phases are vanishing, based on the behavior of the ideal trajectory for the time translation. The following functions satisfy both Eqs. (60) and (66), for example:

$$f(t) = f_0\sin(8\pi\xi t), \quad g(t) = g_0\sin(8\pi\eta t), \tag{67}$$

where f_0 (g_0) is a positive real number and ξ (η) is an integer ($0 \le t \le 1$). In addition, we discuss the values of the dynamical phases when $f(t)$ and $g(t)$ do not satisfy Eq. (66). We choose $f(t)$ and $g(t)$ satisfying Eq. (60) as follows:

$$f(t) = f_0\sin(\pi\xi u_i(t)), \quad g(t) = g_0\sin(\pi\eta u_i(t)), \tag{68}$$

at $t \in [t_{i-1}, t_i]$, where $u_i(t) = (t - t_{i-1})/(t_i - t_{i-1})$. We can show that $\tilde{\gamma}_{\pm d} = 0$ within numerical accuracy if ξ and η are enough large; for example, $\tilde{\gamma}_{+d} = 8.72 \times 10^{-11}$ when $\xi = \eta = 10$ and $f_0 = g_0 = 0.1$.[15] Thus, the dynamical phases can be vanishing when $f(t)$ and $g(t)$ rapidly oscillates around zero.

Now, let us calculate the values of the geometric phases $\tilde{\gamma}_{g\pm}$. We have already shown that there exist $\tilde{\gamma}_\pm \in \mathbb{R}$ such that $|\tilde{n}_\pm(1)\rangle = e^{i\tilde{\gamma}_\pm}|\tilde{n}_\pm\rangle$. When $\tilde{\gamma}_{d\pm} = 0$, the dynamical component of $\tilde{\gamma}_\pm$ is vanishing (i.e., $\tilde{\gamma}_{g\pm} = \tilde{\gamma}_\pm$). We obtain the values of $\tilde{\gamma}_\pm$, solving the Schrödinger equation directly. We choose the corresponding state vector $\boldsymbol{n}_+ = {}^t(0, 1, 0)$ as $|\boldsymbol{n}_+\rangle = e^{i\pi/4}(|0\rangle + i|1\rangle)/\sqrt{2}$. The state vector $|\boldsymbol{n}_+(t)\rangle$ in the presence of the noise is fluctuated around the ideal time evolution by $90_x180_y90_x$, but coincides with it at the final time. We can find that, in both cases, $|\boldsymbol{n}_+\rangle$ evolves to $e^{-i\pi/2}|\boldsymbol{n}_+\rangle$.[15] Therefore, $\tilde{\gamma}_{g+} = -\pi/2$. Furthermore, we find $\tilde{\gamma}_{g-} = +\pi/2$. When we choose the other values of the parameters in Eq. (68) or we use Eq. (67), we can also find the same results: $\tilde{\gamma}_{g\pm} = \mp\pi/2$.

Taking account of the effect under which the time evolution is non-cyclic (e.g., to relax the condition (60)), we can find errors occur, as is pointed out in Ref. 9.[46]

Summarizing the above arguments, the composite pulse $90_x180_y90_x$ is still regarded as a non-adiabatic GQG even in the presence of the fluctuation, if the fluctuation satisfy several conditions. The fluctuations in our model is described by two smooth functions $f(t)$ and $g(t)$. First of all, they have to satisfy Eq. (60) in order to arise well-defined geometric phases. Then, we find three kinds of the conditions for the vanishing dynamical phases: (i) $g(t) = 0$, (ii) $f(t)$ and $g(t)$ have periodicity, or (iii) $f(t)$ and $g(t)$ rapidly oscillates around zero. The composite pulse $90_x180_y90_x$ under infulence of these fluctuations is regarded as a non-adiabatic GQG. From these observations, we expect that the composite pulse $90_x180_y90_x$ is robust against a particular class of noise.

7. Summary

In this lecture note, we showed that the composite pulse $90_x180_y90_x$ is regarded as a non-adiabatic GQG and the condition for the noise which does not affect $90_x180_y90_x$ as a GQG. This condition for the noise is commonly observed in practice and thus we believe that the composite pulse $90_x180_y90_x$ is robust in real experiments. It is interesting that geometric phases, which are the topics on the foundation of quantum mechanics and catch the attention of many researchers in the field of QIP, have been used

144

as a standard technique in the field of NMR.

Acknowledgments

This work was supported by "Open Research Center" Project for Private Universities: matching fund subsidy from MEXT (Ministry of Education, Culture, Sports, Science and Technology).

Appendix A. Solution of Bloch equaiton for a one-qubit system

We briefly show the general solution of the Bloch equation (26). We assume that the time-dependence of $\omega(t)$ and $\boldsymbol{m}(t)$ are given by Eqs. (29)-(31). The Bloch equation is given as

$$\frac{d}{dt}\boldsymbol{n}(t) = \omega(t)\boldsymbol{m}(t) \times \boldsymbol{n}(t). \tag{A.1}$$

When $t \in (t_{i-1}, t_i)$, Eq. (A.1) is reduced to

$$\frac{d}{dt}\boldsymbol{n}(t) = \omega_i \boldsymbol{m}_i \times \boldsymbol{n}(t). \tag{A.2}$$

Here, the corresponding Schrödinger equation is given by $i\frac{d}{dt}|\boldsymbol{n}(t)\rangle = \frac{1}{2}\omega_i \boldsymbol{m}_i \cdot \boldsymbol{\sigma}|\boldsymbol{n}(t)\rangle$. We can readily find that $|\boldsymbol{n}(t)\rangle = e^{-i\theta_i(t)\boldsymbol{m}_i\cdot\boldsymbol{\sigma}/2}|\boldsymbol{n}(t_{i-1})\rangle$, where $\theta_i(t) = \omega_i(t_i - t_{i-1})$ and $t \in [t_{i-1}, t_i)$. According to the correspondence between the state vector $|\boldsymbol{n}(t)\rangle$ and the Bloch vector $\boldsymbol{n}(t)$, we find that

$$\begin{aligned}
\frac{1}{2}\boldsymbol{n}(t) \cdot \boldsymbol{\sigma} &= \frac{1}{2}e^{-i\theta_i(t)\boldsymbol{m}_i\cdot\boldsymbol{\sigma}/2}\,\boldsymbol{n}(t_{i-1}) \cdot \boldsymbol{\sigma}\,e^{+i\theta_i(t)\boldsymbol{m}_i\cdot\boldsymbol{\sigma}/2} \\
&= \frac{1}{2}\Bigg[\cos^2 \frac{\theta_i(t)}{2}\,\boldsymbol{n}(t_{i-1}) \cdot \boldsymbol{\sigma} \\
&\qquad -i\cos\frac{\theta_i(t)}{2}\sin\frac{\theta_i(t)}{2}\,m_{ik}n_l(t_{i-1})[\sigma_k, \sigma_l] \\
&\qquad +\sin^2\frac{\theta_i(t)}{2}\,m_{ij}n_k(t_{i-1})m_{il}\,\sigma_j\sigma_k\sigma_l \Bigg] \\
&= \frac{1}{2}\Bigg[\cos\theta_i(t)\,\boldsymbol{n}(t_{i-1}) \cdot \boldsymbol{\sigma} + \sin\theta_i(t)\,(\boldsymbol{m}_i \times \boldsymbol{n}(t_{i-1})) \cdot \boldsymbol{\sigma} \\
&\qquad +\sin^2\frac{\theta_i(t)}{2}\,(\boldsymbol{m}_i \cdot \boldsymbol{n}(t_{i-1}))\,\boldsymbol{m}_i \cdot \boldsymbol{\sigma} \Bigg].
\end{aligned}$$

It should be noted that we take the summation over the repeating indices (i.e., j, l and k, except for i) in the above expression. Consequently, we have, for $t \in [t_{i-1}, t_i)$,

$$n(t) = \cos\theta_i(t)\, n(t_{i-1}) + \sin\theta_i(t)\,(m_i \times n(t_{i-1}))$$
$$+ (1 - \cos\theta_i(t))\,(m_i \cdot n(t_{i-1}))\, m_i. \qquad (A.3)$$

Remark that the third term in Eq. (A.3) disappears when $m_i \cdot n(t_{i-1}) = 0$, which is equivalent to the vanishing dynamical phase. Accordingly, the solution with no dynamical phase is

$$n(t_i) = \cos\theta_i(t)\, n(t_{i-1}) + \sin\theta_i(t)\,(m_i \times n(t_{i-1})) \quad (t_{i-1} \leq t < t_i). \quad (A.4)$$

Now, we consider a special, but important case. We assume that $\chi_i = 0$ for any i. In this case, we find that $m_i = {}^t(\cos\phi_i, \sin\phi_i, 0)$. It should be noted that one can always express $n(t_{i-1})$ which satisfies $m_i \cdot n(t_{i-1}) = 0$ as

$$n(t_{i-1}) = \begin{pmatrix} \sin\alpha_{i-1}\sin\phi_i \\ -\sin\alpha_{i-1}\cos\phi_i \\ \cos\alpha_{i-1} \end{pmatrix}, \qquad (A.5)$$

where $\alpha_{i-1} \in \mathbb{R}$. Using Eqs. (A.4) and (A.5), we have

$$n(t) = \begin{pmatrix} \sin(\theta_i(t) + \alpha_{i-1})\sin\phi_i \\ -\sin(\theta_i(t) + \alpha_{i-1})\cos\phi_i \\ \cos(\theta_i(t) + \alpha_{i-1}) \end{pmatrix} \quad (t_{i-1} \leq t < t_i). \qquad (A.6)$$

When $i = 0$ (i.e., at the initial time), one can choose α_0 as an arbitrary value. Otherwise, α_i is determined by the recurrence formula,

$$\alpha_i = \theta_i + \alpha_{i-1} \quad (i = 1, 2, \ldots, N), \qquad (A.7)$$

where $\theta_i = \theta_i(t_i) = \omega_i(t_i - t_{i-1})$. Summarizing the above arguments, we have

$$n(t) = \begin{pmatrix} \sin\theta(t)\sin\phi(t) \\ -\sin\theta(t)\cos\phi(t) \\ \cos\theta(t) \end{pmatrix} \quad (t_0 \leq t \leq t_N), \qquad (A.8)$$

where

$$\theta(t) = \sum_{i=1}^{N}(\theta_i(t) + \alpha_{i-1})w(t; t_i, t_{i-1}).$$

Remark that for any i

$$\lim_{t \to t_i+0}\theta(t) = \theta_{i+1}(t_i) = \alpha_i, \quad \lim_{t \to t_i-0}\theta(t) = \theta_i(t_i) = \theta_i + \alpha_{i-1} = \alpha_i.$$

It indicates that $\theta(t)$ is a coninuous function in $[t_0, t_N]$.

Appendix B. Derivation of Hamiltonian with fluctuations

Let us introduce the effect of fluctuations into the path (A.8) as follows:

$$\tilde{n}(t) = \begin{pmatrix} \sin\tilde{\theta}(t)\sin\tilde{\phi}(t) \\ -\sin\tilde{\theta}(t)\cos\tilde{\phi}(t) \\ \cos\tilde{\theta}(t) \end{pmatrix}, \tag{B.1}$$

where

$$\tilde{\theta}(t) = \theta(t) + f(t), \tag{B.2}$$

$$\tilde{\phi}(t) = \phi(t) + g(t). \tag{B.3}$$

The functions $f(t)$ and $g(t)$ are assume to be continuous and smooth functions in $[t_0, t_N]$, and

$$f(t_0) = g(t_0) = 0, \quad f(t_N) = g(t_N) = 0. \tag{B.4}$$

The requirement (B.4) ensures that $\tilde{n}(t)$ is a closed curve on the Bloch sphere even though the fluctuation are introduced.

Let us focus on the time interval $[t_{i-1}, t_i]$. In this domain, we can differenciate $\tilde{\theta}(t)$ and $\tilde{\phi}(t)$ with respect to time. Thus, we find that

$$\frac{d}{dt}\tilde{n}_x = \frac{d\tilde{\theta}}{dt}\cos\tilde{\theta}\sin\tilde{\phi} + \frac{d\tilde{\phi}}{dt}\sin\tilde{\theta}\cos\tilde{\phi},$$

$$\frac{d}{dt}\tilde{n}_y = -\frac{d\tilde{\theta}}{dt}\cos\tilde{\theta}\cos\tilde{\phi} + \frac{d\tilde{\phi}}{dt}\sin\tilde{\theta}\sin\tilde{\phi},$$

$$\frac{d}{dt}\tilde{n}_z = -\frac{d\tilde{\theta}}{dt}\sin\tilde{\theta}.$$

Summarizing the above results, we have

$$\frac{d}{dt}\tilde{n} = (\omega_i + \dot{f}) \begin{pmatrix} \cos\tilde{\phi} \\ \sin\tilde{\phi} \\ 0 \end{pmatrix} \times \begin{pmatrix} \sin\tilde{\theta}\sin\tilde{\phi} \\ -\sin\tilde{\theta}\cos\tilde{\phi} \\ \cos\tilde{\theta} \end{pmatrix}$$

$$+ \dot{g} \begin{pmatrix} 0 \\ 0 \\ 1 \end{pmatrix} \times \begin{pmatrix} \sin\tilde{\theta}\sin\tilde{\phi} \\ -\sin\tilde{\theta}\cos\tilde{\phi} \\ \cos\tilde{\theta} \end{pmatrix}.$$

Therefore, the corresponding Hamiltonian at $t \in [t_{i-1}, t_t]$ is given as

$$\tilde{H}_i(t) = \frac{1}{2}\tilde{\omega}_i(t)\tilde{m}_i(t) \cdot \boldsymbol{\sigma} + \frac{1}{2}\frac{dg(t)}{dt}\sigma_z, \tag{B.5}$$

where

$$\tilde{\omega}_i(t) = \omega_i + \frac{df(t)}{dt},$$

$$\tilde{\boldsymbol{m}}_i(t) = \begin{pmatrix} \cos(\phi_i + g(t)) \\ \sin(\phi_i + g(t)) \\ 0 \end{pmatrix}.$$

Consequently, we find that the Hamiltonian which describes the fluctuated system is given by

$$\tilde{H}(t) = \frac{1}{2}\tilde{\omega}(t)\tilde{\boldsymbol{m}}(t) \cdot \boldsymbol{\sigma} + \frac{1}{2}\frac{dg(t)}{dt}\sigma_z, \tag{B.6}$$

where

$$\tilde{\omega}(t) = \sum_{i=1}^{N} \tilde{\omega}_i(t)w(t; t_i, , t_{i-1}), \tag{B.7}$$

$$\tilde{\boldsymbol{m}}(t) = \sum_{i=1}^{N} \tilde{\boldsymbol{m}}_i(t)w(t; t_i, t_{i-1}). \tag{B.8}$$

References

1. J. A. Jones, V. Vedral, A. Ekert and G. Castagnoli: Nature **403** (2000) 869.
2. A. Ekert, M. Ericsson, P. Hayden, H. Inamori, J. A. Jones, D. K. L. Oi and V. Vedral: J. Mod. Opt. **47** (2000) 2501.
3. S.-L. Zhu and Z. D. Wang: Phys. Rev. A **67** (2003) 022319.
4. M. Tian, Z. W. Barber, J. A. Fischer and Wm. Randall Babbitt: Phys. Rev. A **69** (2004) 050301(R).
5. R. Das, S. K. K. Kumar and A. Kumar: J. Magn. Reson. **177** (2005) 318.
6. H. Imai and A. Morinaga: Phys. Rev. A **76** (2007) 062111.
7. Y. Ota, Y. Goto, Y. Kondo and M. Nakahara: arXiv:0906.0474.
8. A. Nazir, T. P. Spiller and W. J. Munro: Phys. Rev. A **65** (2002) 042303.
9. A. Blais and A.-M. S. Tremblay: Phys. Rev. A **67** (2003) 012308.
10. G. De Chiara and G. M. Palma: Phys. Rev. Lett. **91** (2003) 090404.
11. A. Carollo, I. Fuentes-Guridi, M. F. Santos and V. Vedral: Phys. Rev. Lett. **92** (2004) 020402.
12. S.-L. Zhu and P. Zanardi: Phys. Rev. A **72** (2005) 020301(R).
13. G. De Chiara, A. Łoziński and G. M. Palma: Eur. Phys. J. D **41** (2007) 179.
14. J. Dajka, M. Mierzejewski and J. Łuczka: J. Phys. A: Math. Theor. **41** (2008) 012001.
15. Y. Ota and Y. Kondo: Phys. Rev. A **80** (2009) 024302.
16. M. H. Levit: Prog. Nuc. Magn. Reson. Spectrose. **18** (1986) 61.
17. R. R. Ernst, G. Bodenhausen and A. Wokaun: *Principles of Nuclear Magnetic Resonance in One and Two Dimensions* (Oxford University Press, New York, 1987).

18. R. Freeman: *A Handbook of Nuclear Magnetic Resonance* (Longman Scientific and Technical, England, 1988).
19. T. D. W. Claridge: *High-Resolution NMR Techniques in Organic Chemistry* (Pergamon, Amsterdam, 1999).
20. Y. Aharonov and J. Anandan: Phys. Rev. Lett. **58** (1987) 1593.
21. M. Nakahara and T. Ohmi: *Quantum Computing: From Linear Algebra To Physical Realizations* (CRC Press, New York, 2008).
22. O. Chuluunbaatar, V. L. Derbov, A. Galtbyar, A. A. Gusev, M. S. Kaschiev, S. I. Vinitsky and T. Zhanlav: J. Phys. A:Math. Theor. **41** (2008) 295203.
23. H. K. Cummins and J. A. Jones: New J. Phys. **2** (2000) 6.
24. H. K. Cummins, G. Llewellyn and J. A. Jones: Phys. Rev. A **67** (2003) 042308.
25. Z. Zhou, S.-I. Chu and S. Han: Phys. Rev. B **66** (2002) 054527.
26. M. Steffen, J. M. Martinis and I. L. Chuang: Phys. Rev. B **68** (2003) 224518.
27. T. Henage, M. Delaney, E. Urban, T. Johnson, L. Isenhower, D. Yavuz, T. Walker and M. Saffman: DAMOP07 Meeting of the American Physical Society (2007) 5.3(E).
28. R. Shewmon, J. Labaziewicz, Y. Ge, S. Wang and I. L. Chuang: MAR08 Meeting of the American Physical Society (2008) 23.3(E).
29. M. Berry: Proc. R. Soc. Lond. A **392** (1984) 45.
30. B. Simon: Phys. Rev. Lett. **51** (1983) 2167.
31. M. Nakahara: *Geometry, Topology and Physics* (IOP Publishing, Bristol and Philadelphia, 2003).
32. D. N. Page: Phys. Rev. A **36** (1987) 3479.
33. J. Samuel and R. Bhandari: Phys. Rev. Lett. **60** (1988) 2339.
34. S. Pancharatnam: Proc. Ind. Acad. Sci. A **44** (1956) 247.
35. M. V. Berry: J. Mod. Opt. **34** (1987) 1401.
36. S.-L. Zhu, Z. D. Wang and Y.-D. Zhang: Phys Rev. B **61** (2000) 1142.
37. A. Uhlmann: Rep. Math. Phys. **24** (1986) 229.
38. A. Uhlmann: Lett. Math. Phys. **21** (1991) 229.
39. E. Sjöqvist, A. K. Pati, A. Ekert, J. S. Anandan, M. Ericsson, D. K. L. Oi and V. Vedral: Phys. Rev. Lett. **85** (2000) 2845.
40. K. Singh, D. M. Tong, K. Basu, J. L. Chen and J. F. Du: Phys. Rev. A **67** (2003) 032106.
41. J. Du, P. Zou, M. Shi, L. C. Kwek, J.-W. Pan, C. H. Oh, A. Ekert, D. K. L. Oi and M. Ericsson: Phys. Rev. Lett. **91** (2003) 100403.
42. D. M. Tong, E. Sjöqvist, L. C. Kwek and C. H. Oh: Phys. Rev. Lett. **93** (2004) 080405.
43. D. Chruściński and A. Jamiołkowski: *Geometric Phases in Classical and Quantum Mechanics* (Birkhäuser, Boston, 2004).
44. I. Bengtsson and K. Życzkowski: *Geometry of Quantum States: An Introduction to Quantum Entanglement* (Cambridge University Press, New York, 2006).
45. D. Suter, K. T. Mueller and A. Pines: Phys. Rev. Lett. **60** (1988) 1218.

46. It is interesting to consider a supercycle.[16,19] We can numerically find that the supercycles based on $90_x 180_y 90_x$ approximately correspond to the cyclic evolutions. It means that one may find a geometric phase in the supercycles.

QUANTUM WIPE EFFECT

AKIRA SAITOH[1], ROBABEH RAHIMI[1], MIKIO NAKAHARA[1,2]

[1] *Research Center of Quantum Computing, Interdisciplinary Graduate School of Science and Engineering, Kinki University, 3-4-1 Kowakae, Higashi-Osaka, Osaka 577-8502, Japan*
[2] *Department of Physics, Kinki University, 3-4-1 Kowakae, Higashi-Osaka, Osaka 577-8502, Japan*

We consider a model of a spin system under the influence of decoherence such that a system coupled with a dissipating environmental system consisting of either spins or bosonic modes. The dissipation of an environment is governed by a certain probability with which an environmental system localized around a principal system dissipates into a larger bath and a thermal environmental system instead migrates into the place. A certain threshold on the probability is found in the growth of decoherence in a principal system. A larger as well as a smaller dissipation probability than the threshold results in smaller decoherence.

This finding is utilized to elucidate a spin relaxation theory of a magnetic resonance spectrometer. In particular, a seamless description of transverse relaxation and motional narrowing is possible.

We also numerically evaluate the dynamics of coherence useful for quantum information processing. The bang-bang control and anti-Zeno effect in entanglement and the Oppenheim-Horodecki nonclassical correlation are investigated in the model of spin-boson coupling.

Keywords: Decoherence; Nonclassical Correlation; Motional Narrowing

1. Introduction

The system of our interest is a spin system coupled with an external spin bath or an external bosonic bath. Spin relaxation has been commonly studied in magnetic resonance theories using a phenomenological rate equation.[1] Nevertheless, time evolutions of quantum coherence with phase factors cannot be traced by the phenomenological theory.

The model we introduce here is simple but more plausible than the rate equation. It has a principal spin system and an environmental system evolving under a certain Hamiltonian. With a certain probability, the environmental system dissipates into a larger bath system and a thermal environmental system migrates into the place. We utilize a recurrence formula to solve this model either analytically or numerically.

We further study simple dynamical and static control of decoherence in spin systems. In particular, the bang-bang control[2,3] and the quantum wipe effect[4-7] are applied for suppressing decoherence. Bang-bang control uses regular short pulses to decouple spins from environmental systems; it is commonly known as basic dynamical decoupling in nuclear magnetic resonance (NMR) spectroscopy.[8] Quantum wipe effect is a reduction of decoherence caused by an increase of environmental dissipation rate over a certain threshold. It might be referred to as motional narrowing known in NMR[9,10] although the formulation has been different so far.

The dynamics of quantum coherence[11,12] has been extensively studied in the context of quantum information processing.[13,14] Advantages over classical information processing, such as unconditional security of communication,[15] exponential speedup of computation,[16,17] etc., are considered to be owing to quantum coherence. For quantum information processing using multipartite systems, quantumness in correlations is the common resource to achieve a nonclassical functionality. This includes quantum teleportation[18] and its variants[19-24] as well as quantum data hiding,[25-27] quantum secret sharing,[28,29] etc. These protocols require entangled pure states in normal setting although it is known that quantum data hiding can be performed with quantumness without entanglement.[26] Thus the amount of entanglement remaining in a time evolution is a good measure of robustness of quantum processing against noise.

Entanglement is also known as a resource to perform quantum computation that cannot be easily simulated by classical computers. In fact, it is known[30] that any quantum computation using pure states whose Schmidt rank grows polynomially in the number of qubits can be efficiently simulated by classical computers. It is, however, not known if a mixed state with small or vanishing entanglement can be simulated efficiently. A purification of a mixed state with small or vanishing entanglement often possesses a large Schmidt rank.

A typical algorithm working without entanglement is Knill and Laflamme's fast estimation of a normalized trace of a unitary matrix, which uses a single pseudo-pure qubit and the remaining qubits in a maximally mixed state.[31] Recently, Datta et $al.$[32] revealed that it involves a large quantum discord[33] and vanishing entanglement. Quantum discord is a distance of two different expressions of mutual information that should be identical in a classical probabilistic theory. It is one of the measures of quantumness in correlation.[33-36]

Adherence to the entanglement paradigm often causes a misunderstanding that no more quantum processing is possible under the lack of entanglement. As a typical example, entanglement sudden death[37] sounds like a sudden death of quantumness in processing. Nevertheless, correct discussions should involve evaluations of quantumness other than entanglement. For example, a pseudo (PS) pure state $\rho_{\mathrm{PS}} = p|\psi\rangle\langle\psi| + (1-p)I/4$ with $|\psi\rangle = (|00\rangle + |11\rangle)/\sqrt{2}$ is separable for $p < 1/3$. Setting $p \sim \exp(-t/\tau)$, with t time and τ the characteristic time, results in the quick death of entanglement at $t = \log 3/\tau$. This is a sort of depolarization effects. The coherence terms [$(00, 11)$ and $(11, 00)$ elements], however, survives during finite t. The identity factor in ρ_{PS} does no effect on quantum processing pertaining to $|\psi\rangle\langle\psi|$ except for the effect of decreasing the signal intensity.

Decoherence harmful to quantum information processing is the one changing the output to unexpected one. In this sense, loss of entanglement is not essentially important. Thus the robustness of quantum processing against noise is more reliably measured if quantumness in correlation is traced in addition to entanglement.

In this paper, we investigate the loss of nonclassical correlation in realistic physical models. Not only entanglement but the zero-way quantum deficit[35] are evaluated in numerical simulations of time-evolving spin systems. It is expected that the zero-way quantum deficit survives longer than entanglement.

This paper is organized as follows.

First we will briefly revisit conventional phenomenological theories describing a transverse relaxation in Secs. 2.1 and 2.2. The theory in Sec. 2.2 is explained with a slight modification for improving an accuracy of a rate equation. We also briefly revisit a motional narrowing theory in a classical postulate in Sec. 2.3.

We then introduce a simple model to describe the transverse relaxation of a spin 1/2 in Sec. 3. We show a derivation of decoherence in the off-diagonal elements of a density matrix of a principal spin system without using correlation functions in Sec. 3.1. The derivation involves recurrence formula pertaining to a time evolution of the density matrix.

In Sec. 4, we further numerically evaluate the decoherence of a single spin 1/2 coupled to a on-resonance bosonic mode and that of a two-spin 1/2 system couple to on-resonance bosonic modes. Time evolutions of zero-way quantum deficit and entanglement are investigated.

The paper ends with the concluding remarks in Sec. 5.

2. Historical Model for Transverse Relaxation

2.1. *Phenomenological description for* T_2 *time relaxation*

A spin ensemble of our interest is an ensemble of physically identical spin systems that are individually in certain states. The state of the ensemble is an average of the individual states over all the spin systems. An ensemble is described by a density matrix ρ and its magnetization in a certain direction (say, $d = x, y, z$) is proportional to $M_d \equiv \mathrm{Tr} M_d \rho$ with M_d a certain operator associated to a spacial direction (e.g., Pauli operators). A transverse relaxation is a decay of a transverse magnetization of a spin ensemble. For a single-spin 1/2 system, the magnetization in x direction is zero at a thermal equilibrium under a strong static magnetic field incurred in the z direction. To observe a transverse relaxation, one may title the z-directional magnetization to the horizontal plane.

Let the magnetization in the x direction be $M_x(t)$. Suppose it has the largest value at $t = 0$. Then, it is a convention[38,39] to consider the relaxation governed by

$$\frac{\partial M_x}{\partial t} = -\frac{M_x}{T_2}.$$

with the transverse (T_2) relaxation time T_2. A solution is easily obtained as $M_x(t) = M_x(0) \exp(-t/T_2)$.

Experimentally observed T_2 relaxations in magnetic resonance spectrometries like NMR or ESR are usually consistent to this phenomenological equation. One drawback of the equation is that we cannot describe a more complicated phenomena like motional narrowing. In a conventional theory, an additional phenomenological theory is introduced to explain the motional narrowing as we will see in Sec. 2.3. As we mentioned in the introduction, we will later introduce a model with which one can describe a transverse relaxation and a motional narrowing simultaneously.

A transverse relaxation time is typically observed as a line broadening of peaks in Fourier spectra in magnetic resonance spectroscopy. Let us briefly revisit the relation between T_2 and the broadening.

2.1.1. *FID spectrum and* T_2

A line shape of a spectrum in magnetic resonance spectroscopy is commonly a Fourier transformation of a free time evolving transverse magnetization of a spin ensemble after the magnetization is titled onto the horizontal plane. It is called free induction decay (FID) spectrum. The ideal transverse magnetization for a spin-1/2 system under the FID is given by

$$M_x(t) = M_x(0)e^{-t/\mathrm{T_2}}\sin(\omega_L t)$$

with ω_L a Larmor frequency of the precession of the spin.

Usually one records $M_x(t)$ and $M_y(t)$ with the phase difference of $\pi/2$. Two records are mixed and we obtain the following data $D(t)$.

$$\begin{aligned}D(t) &= M_x(t) - iM_y(t)\\ &= M_x(0)e^{-t/\mathrm{T_2}}\left[i\sin(\omega_L t) + \cos(\omega_L t)\right]\\ &= e^{-t/\mathrm{T_2}}\exp(i\omega_L t).\end{aligned}$$

Commonly, the FID spectrum is the Fourier transformation of $D(t)$.

To compute a Fourier spectrum, usually the envelope $e^{-t/\mathrm{T_2}}$ is modified to be a reflection-symmetric shape:

$$\begin{cases}\exp(t/\mathrm{T_2}) & t < 0\\ \exp(-t/\mathrm{T_2}) & t \geq 0\end{cases}$$

This results in the amplitude of the Fourier spectrum

$$\begin{aligned}\mathcal{F}[D(t)] &= M_x(0)\int_{-\infty}^{\infty} g(t)\exp[-i(\omega - \omega_L)t]\,dt\\ &= M_x(0)\left[\frac{2\mathrm{T_2}}{1 + \mathrm{T_2^2}(\omega - \omega_L)^2} + \text{const}\right].\end{aligned}$$

The constant const should be truncated. The shape of $\mathcal{F}[D(t)]$ is depicted as illustrated in Fig. 1. The value of $\mathrm{T_2}$ is a reciprocal of a half width at half maximum of the peak.

We have revisited the idealized relation between a line shape of an FID spectrum and $\mathrm{T_2}$. Real experimental results should be analyzed by a more realistic model involving chemical shifts, hyperfine coupling, J coupling, dipole-dipole coupling, inhomogeneity of magnetic fields, errors in calibration, etc. The derivation of $\mathrm{T_2}$ relaxation, described as an exponential decay, due to a spin-spin coupling is conventionally performed on the basis of the Bloembergen-Purcell-Pound (BPP) theory as we overview in the next subsection. After the revisit to the BPP theory, we will introduce our model for the derivation from a different approach in Sec. 3.

2.2. *Transverse relaxation in a modified BPP theory*

The Bloembergen-Purcell-Pound (BPP) theory[1] is a standard phenomenological approach for a transverse ($\mathrm{T_2}$) relaxation.[8,40] The theory is, however,

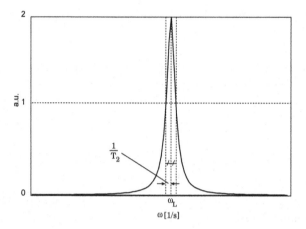

Fig. 1. Illustration of $\mathcal{F}[D(t)]$. The half width at half maximum of the peak is $1/\mathrm{T}_2$.

not accurate in the initial approximation and results in a nonaccurate representation of T_2 as a function of transition probabilities. Here, we follow the BPP theory with a slight modification so that we can achieve a more accurate representation of T_2.

Recall here that T_2 is a characteristic time in the process of losing magnetization of a spin $(1/2)$ in the x direction. Following conventions, we assume that T_2 relaxation is mainly due to a spin-spin coupling between a principal spin I and an environmental spin S, and relaxation can be analyzed using probabilities of transitions among eigenstates of $I_x \otimes \mathbb{1}$ and $\mathbb{1} \otimes S_x$.

Let us use the vectors

$$|\widetilde{++}\rangle = |00\rangle + |11\rangle + |01\rangle + |10\rangle,$$
$$|\widetilde{+-}\rangle = |00\rangle - |11\rangle - |01\rangle + |10\rangle,$$
$$|\widetilde{-+}\rangle = |00\rangle - |11\rangle + |01\rangle - |10\rangle,$$
$$|\widetilde{--}\rangle = |00\rangle + |11\rangle - |01\rangle - |10\rangle,$$

where $|0\rangle$ and $|1\rangle$ are eigenvectors of I_z (and S_z). These four vectors are the eigenstates of $I_x \otimes \mathbb{1}$ and also of $\mathbb{1} \otimes S_x$. Let us also follow the convention to use transition probabilities, u_0, u_1, u_2, u_1', for transitions between these states in a classical manner, for a certain time duration τ [sec], as illustrated in the diagram shown in Fig. 2. The populations of these states are represented as $N_{\widetilde{++}}, N_{\widetilde{+-}}, N_{\widetilde{-+}}, N_{\widetilde{--}}$.

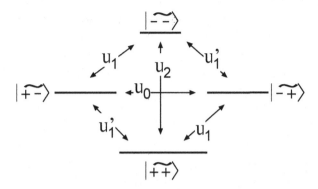

Fig. 2. Transition probabilities between the eigenstates.

The conventional BPP theory uses the rate equation given as follows.

$$\tau\frac{\partial N_{\widetilde{++}}}{\partial t} = -(u_1 + u_1' + u_2)N_{\widetilde{++}} + u_1'N_{\widetilde{+-}} + u_1 N_{\widetilde{-+}}$$
$$+ u_2 N_{\widetilde{--}} + c_{\widetilde{++}},$$
$$\tau\frac{\partial N_{\widetilde{+-}}}{\partial t} = u_1'N_{\widetilde{++}} - (u_0 + u_1 + u_1')N_{\widetilde{+-}} + u_0 N_{\widetilde{-+}}$$
$$+ u_1 N_{\widetilde{--}} + c_{\widetilde{+-}},$$
$$\tau\frac{\partial N_{\widetilde{-+}}}{\partial t} = u_1 N_{\widetilde{++}} + u_0 N_{\widetilde{+-}} - (u_0 + u_1 + u_1')N_{\widetilde{-+}}$$
$$+ u_1'N_{\widetilde{--}} + c_{\widetilde{-+}},$$
$$\tau\frac{\partial N_{\widetilde{--}}}{\partial t} = u_2 N_{\widetilde{++}} + u_1 N_{\widetilde{+-}} + u_1'N_{\widetilde{-+}}$$
$$- (u_1 + u_1' + u_2)N_{\widetilde{--}} + c_{\widetilde{--}},$$

(1)

where $c_{\widetilde{\pm\pm}}$ are constants describing interactions between the spin and a certain lattice. These equations seem natural at a glance, but not accurate. For example, the first equation can be rewritten as

$$\tau[N_{\widetilde{++}}(t + \Delta t) - N_{\widetilde{++}}(t)]$$
$$= -(u_1 + u_1' + u_2)\Delta t\ N_{\widetilde{++}}(t) + u_1'\Delta t\ N_{\widetilde{+-}}(t)$$
$$+ u_1\Delta t\ N_{\widetilde{-+}}(t) + u_2\Delta t N_{\widetilde{--}}(t) + c_{\widetilde{++}}\Delta t.$$

In this equation, $u_1\Delta t/\tau$ etc. is assumed to be a transition probability for a duration Δt. This is not an accurate description.

158

In a more accurate description, the transition probability for the transition from $|\widetilde{++}\rangle$ to $|\widetilde{-+}\rangle$ for the finite duration Δt should be written as

$$1 - (1 - u_1)^{\Delta t/\tau}.$$

As Δt is small, one may approximate it as

$$-\frac{\Delta t}{\tau} \log(1 - u_1).$$

We replace u_0, u_1, $u_1{}'$ and u_2 with $-\log(1-u_0)$ $-\log(1-u_1)$, $-\log(1-u_1{}')$, and $-\log(1-u_2)$, respectively, in Eq. (1). For simplicity, let us define

$$\begin{aligned}
\hat{u}_0 &\equiv -\log(1 - u_0), \quad \hat{u}_1 \equiv -\log(1 - u_1),\\
\hat{u}_1' &\equiv -\log(1 - u_1'), \quad \text{and} \quad \hat{u}_2 \equiv -\log(1 - u_2).
\end{aligned} \tag{2}$$

In addition to the rate equation, we have the equations describing the average magnetization in the x direction:

$$\begin{aligned}
K_I \overline{< I_x >} &= N_{\widetilde{++}} + N_{\widetilde{+-}} - N_{\widetilde{-+}} - N_{\widetilde{--}},\\
K_S \overline{< S_x >} &= N_{\widetilde{++}} + N_{\widetilde{-+}} - N_{\widetilde{+-}} - N_{\widetilde{--}},
\end{aligned} \tag{3}$$

where K_I and K_S are proportionality constants.

We obtain the following equations from Eq. (1) with modification (2) and Eq. (3).

$$\begin{aligned}
\tau K_I \frac{\partial \overline{< I_x >}}{\partial t} &= -2(\hat{u}_1 + \hat{u}_2)(N_{\widetilde{++}} - N_{\widetilde{--}})\\
&\quad -2(\hat{u}_1 + \hat{u}_0)(N_{\widetilde{+-}} - N_{\widetilde{-+}}) + c_1,\\
\tau K_S \frac{\partial \overline{< S_x >}}{\partial t} &= -2(\hat{u}_1' + \hat{u}_2)(N_{\widetilde{++}} - N_{\widetilde{--}})\\
&\quad +2(\hat{u}_1' + \hat{u}_0)(N_{\widetilde{+-}} - N_{\widetilde{-+}}) + c_2,
\end{aligned} \tag{4}$$

with $c_1 = c_{\widetilde{++}} + c_{\widetilde{+-}} - c_{\widetilde{-+}} - c_{\widetilde{--}}$ and $c_1 = c_{\widetilde{++}} - c_{\widetilde{+-}} + c_{\widetilde{-+}} - c_{\widetilde{--}}$. We also obtain the following equations from Eq. (3).

$$\begin{aligned}
K_I \overline{< I_x >} + K_S \overline{< S_x >} &= 2(N_{\widetilde{++}} - N_{\widetilde{--}}),\\
K_I \overline{< I_x >} - K_S \overline{< S_x >} &= 2(N_{\widetilde{+-}} - N_{\widetilde{-+}}).
\end{aligned} \tag{5}$$

Equations (4) and (5) lead to

$$\tau\frac{\partial\overline{< I_x >}}{\partial t} = -(\hat{u}_0 + 2\hat{u}_1 + \hat{u}_2)\overline{< I_x >}$$
$$- (\hat{u}_2 - \hat{u}_0)\frac{K_S}{K_I}\overline{< S_x >} + \frac{c_1}{K_I},$$
$$\tau\frac{\partial\overline{< S_x >}}{\partial t} = -(\hat{u}_0 + 2\hat{u}_1' + \hat{u}_2)\overline{< S_x >}$$
$$- (\hat{u}_2 - \hat{u}_0)\frac{K_I}{K_S}\overline{< I_x >} + \frac{c_2}{K_S}.$$

(6)

It is a convention to assume that $\overline{< I_x >}$ ($\overline{< S_x >}$) vanishes at a certain time of an equilibrium point where $\frac{\partial\overline{< I_x >}}{\partial t} = 0$ ($\frac{\partial\overline{< S_x >}}{\partial t} = 0$). This implies $c_1 = c_2 = 0$. Thus the system of equations consequently becomes

$$\tau\frac{\partial\overline{< I_x >}}{\partial t} = -(\hat{u}_0 + 2\hat{u}_1 + \hat{u}_2)\overline{< I_x >} - (\hat{u}_2 - \hat{u}_0)\frac{K_S}{K_I}\overline{< S_x >},$$
$$\tau\frac{\partial\overline{< S_x >}}{\partial t} = -(\hat{u}_0 + 2\hat{u}_1' + \hat{u}_2)\overline{< S_x >} - (\hat{u}_2 - \hat{u}_0)\frac{K_I}{K_S}\overline{< I_x >}.$$

(7)

T_2 when $I = S$ In case of $I = S$, we can assume that $u_1 = u_1'$ (hence $\hat{u}_1 = \hat{u}_1'$), $\overline{< I_x >} = \overline{< S_x >}$ and $K_I = K_S$. Then we have

$$\tau\frac{\partial\overline{< I_x >}}{\partial t} = -2(\hat{u}_1 + \hat{u}_2)\overline{< I_x >}.$$

The solution of this equation is

$$\overline{< I_x >}(t) = \overline{< I_x >}(0)\exp[-2(\hat{u}_1 + \hat{u}_2)t/\tau].$$

Recall that magnetization $\mathbf{M}$ is given by $N\gamma\hbar\mathbf{I}$ with N the number of $\mathbf{I}$ spins in an ensemble and γ the gyromagnetic ratio. Therefore,

$$M_x(t) = M_x(0)\exp[-2(\hat{u}_1 + \hat{u}_2)t/\tau].$$

Thereby the transverse relaxation time is

$$T_2 = \frac{\tau}{2(\hat{u}_1 + \hat{u}_2)} = \frac{\tau}{-2\log[(1 - u_1)(1 - u_2)]}.$$

(8)

Values of u_1 and u_2 are estimated by using a magnetic filed intensity and intensities of Fourier spectra that are calculated from statistics of the motions of spins.[1]

In the limit of $u_1, u_2 \to 0$, this result converges to the result of the standard BPP theory, $T_2 = \tau/(2u_1 + 2u_2)$. The problem of the original theory was that T_2 much smaller than τ cannot be obtained by setting u_1 and u_2 close to one. Indeed, a realistic relaxation time $T_2 \sim 100$ms, often observed in experiments, can be obtained by setting a proper value to τ. It is, however, unnatural that the probabilities are not the parameters to touch in order for achieving a physically plausible relaxation time.

In contrast to the original BPP theory, our slight modified theory led to an appropriate relaxation time given by Eq. (8). In this equation, u_1 and/or u_2 very close to one result in a small T_2. It is now possible to obtain a realistic $T_2 \sim$ from 10μs to 100ms by setting proper values for u_1 and u_2 with the value of τ fixed. Thus we may employ $\tau = 1$ to reduce the number of parameters.

In addition, it is clear that the theory is, so far, still too simple to explain the phenomenon of motional narrowing. We will see a conventional phenomenological approach in Sec. 2.3 briefly. We will later introduce a model with which a motional narrowing is elucidated naturally together with a transverse relaxation.

2.3. Motional narrowing from a phenomenological approach

It is commonly known in the community of spectroscopy that increasing temperature of a sample tube results in a sharper peak in a spectrum in magnetic resonance spectroscopy. This phenomenon is called motional narrowing.

A common conventional model to elucidate the motional narrowing is the Anderson-Weiss model.[9] It introduces a phenomenological autocorrelation function of a spin precession (or Larmor) frequency of an ensemble. Another model is Kubo-Tomita's quantum statistical one.[10] The quantum statistical model was further developed and the temperature dependence of the narrowing was analytically calculated from approximated Liouville-von Neumann equations.[41,42] Here we briefly revisit the Anderson-Weiss model.

Let us denote the Larmor frequency of a microscopic spin as ω_L. In the Anderson-Weiss model, the typical autocorrelation function of ω_L is assumed to be

$$< \omega_L(t)\omega_L(t - \tau) >=< \omega_L^2 > \exp(-|\tau|/\tau_c)$$

with τ_c a characteristic (or correlation) time. As the Larmor frequency varies time to time, the envelope of an FID (as a function of time t) is given by

$$F(t) = \langle e^{i\kappa(t)} \rangle_{\omega_L} \equiv \int_{-\infty}^{\infty} D(\omega_L)e^{i\kappa(t)}d\omega_L$$

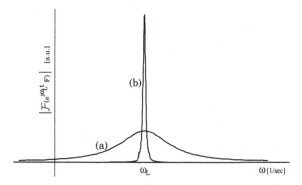

Fig. 3. A crude illustration of the line shape $|\mathcal{F}(e^{i\omega_{\text{L}}t}F)|$. (a) The line shape for large τ_{c}. (b) that for small τ_{c}.

with

$$\kappa(t) = \int_0^t \omega_{\text{L}}(\tau)d\tau$$

and $D(\omega_{\text{L}})$ a distribution in the frequency domain. This leads to

$$F(t) \sim 1 + \frac{\langle\kappa(t)\rangle_{\omega_{\text{L}}}}{1!} + \frac{\langle\kappa(t)^2\rangle_{\omega_{\text{L}}}}{2!} + \cdots$$

$$\simeq e^{-\langle\kappa(t)^2\rangle_{\omega_{\text{L}}}/2}$$

$$= \exp[-\int_0^t (t-\tau) < \omega_{\text{L}}(t)\omega_{\text{L}}(t-\tau) > d\tau].$$

Now we have

$$F(t) \simeq \exp\left[- <\omega_{\text{L}}^2> \int_0^t (t-\tau)\exp(-|\tau|/\tau_{\text{c}})\,d\tau\right]$$

$$= \exp\left\{- <\omega_{\text{L}}^2> \tau_{\text{c}}^2[e^{-|t|/\tau_{\text{c}}} - 1 + |t|/\tau_{\text{c}}]\right\}.$$

Consequently, the FID spectrum is obtained as

$$|\mathcal{F}(e^{i\omega_{\text{L}}t}F)| \simeq \begin{cases} \delta(\omega - \omega_{\text{L}}) & \tau_{\text{c}} \ll 1 \\ \left|\text{const} - \dfrac{2\tau_{\text{c}}\log A}{A\log^2 A + \tau_{\text{c}}^2(\omega-\omega_{\text{L}})^2 A}\right| & \tau_{\text{c}} \gg 1 \end{cases},$$

where $A = \exp(- <\omega_{\text{L}}^2> \tau_{\text{c}}^2)$. This result suggests that the line shape is Gaussian for a large correlation time (i.e., for a slow molecular motion) and it exhibits a very sharp line shape for a small correlation time (i.e., for a rapid molecular motion) as illustrated in Fig. 3.

In this section, a conventional approach to theoretical understanding of motional narrowing has been explained. It is purely classical as is clear from the fact that density matrices are not involved. A drawback is that this theory is conceptually independent of the transverse relaxation theory that we have seen in Sec. 2.2. In the next section, we will introduce a simple semi-classical model with which one can explain transverse relaxation and motional narrowing seamlessly.

3. Semi-classical Model for Transverse Relaxation and Motional Narrowing

We have seen crude reviews on conventional phenomenological approaches to the BPP transverse relaxation theory and the Anderson-Weiss motional narrowing theory. Here, we propose a simple semi-classical model to handle the both phenomena.

Let us assume that the entire environment is so large that the environmental system (system E) coupled with the principal system (system S) is a part of a large environment and hence it is replaced with a thermal environmental system with probability p (namely, with some dissipation rate) per certain time interval τ. Systems S and E are represented by the density matrix ρ^{SE}; a thermal environmental system is represented by the density matrix σ. The Hamiltonian affecting the time evolution is reduced to the one consisting only of the time-independent Hamiltonian H that governs systems S and E including their interaction. This model is illustrated in Fig. 4. For a small time interval Δt, the evolution of the systems S and E obeys the equation

$$\rho^{\mathrm{SE}}(\tilde{t} + \Delta t) = e^{-iH\Delta t}\left[x^{\Delta t}\rho^{\mathrm{SE}}(\tilde{t}) + (1 - x^{\Delta t})\mathrm{Tr}_{\mathrm{E}}\rho^{\mathrm{SE}}(\tilde{t}) \otimes \sigma\right]e^{iH\Delta t}, \quad (9)$$

where $x = (1 - p)^{1/\tau}$ and $\tilde{t}$ denotes a certain time step.

3.1. Theoretical solution for relaxation due to simple spin-spin coupling

To evaluate the T_2 relaxation, let us consider the case where the systems S and E are single-qubit systems. In addition, we impose the following conditions. The principal system is originally represented by a density matrix

$$\rho^{\mathrm{S}}(0) = \begin{pmatrix} a & b \\ b^* & 1 - a \end{pmatrix}$$

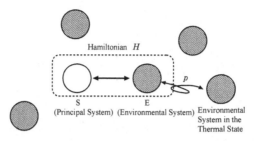

Fig. 4. Model of a system consisting of the principal system (system S) and the environmental system (system E) whose time evolution is governed by the Hamiltonian H. System E is replaced with a thermal environmental system with the dissipation probability p per time interval τ.

with $0 \leq a \leq 1$ and $0 \leq |b| \leq \sqrt{a(1-a)}$. The environmental system at thermal equilibrium is represented by the maximally-mixed density matrix

$$\rho^{E}(0) = \sigma = \begin{pmatrix} 1/2 & 0 \\ 0 & 1/2 \end{pmatrix}.$$

The initial state of the total system is set to $\rho^{SE}(0) = \rho^{S}(0) \otimes \rho^{E}(0)$. The Hamiltonian H is set to

$$cI_z \otimes I_z = \mathrm{diag}(c/4, -c/4, -c/4, c/4), \qquad (10)$$

where $I_z = \mathrm{diag}(1/2, -1/2)$. The conditions imposed here is found in certain physical systems. A typical example is inter-molecular spin-spin coupling between a diluted guest molecule and a host molecule in a liquid sample of an NMR spectrometer. The principal system is a spin 1/2 of the guest molecule and the localized environmental system is the one in a host molecule which is rapidly replaced with one in a bath of the host molecules. The above conditions are reasonable for the case where the gyromagnetic ratio of the spin of the guest molecule is much larger than that of the host molecule whose spin polarization is negligibly small. We should, however, note that the coupling governed by Eq. (10) can be rather weak in comparison to dipole-dipole coupling. A realistic analysis considering many of the possible couplings should wait for future researches. Here we have chosen the setting amenable to an analytical approach.

Let us begin the analysis. With the above simplifications, one can find the form of the density matrix at time $t = m\Delta t$ $(m \in \{0, 1, 2, \ldots\})$ as

$$\rho^{\mathrm{SE}}(m\Delta t) = \begin{pmatrix} a/2 & 0 & f_m & 0 \\ 0 & a/2 & 0 & g_m \\ f_m^* & 0 & (1-a)/2 & 0 \\ 0 & g_m^* & 0 & (1-a)/2 \end{pmatrix},$$

with functions f_m and g_m depending on m. They obey the system of recurrence formulae:

$$\begin{cases} f_{m+1} = \frac{1}{2}e^{-ic\Delta t/2}\left[f_m + g_m + x^{\Delta t}(f_m - g_m)\right] \\ g_{m+1} = \frac{1}{2}e^{ic\Delta t/2}\left[f_m + g_m - x^{\Delta t}(f_m - g_m)\right] \end{cases}$$

with $f_0 = g_0 = b/2$. This leads to the following recurrence formula:

$$\kappa_{m+2} = (1 + x^{\Delta t})\cos(c\Delta t/2)\kappa_{m+1} - x^{\Delta t}\kappa_m, \tag{11}$$

where $\kappa_m = f_m$ or g_m with $f_0 = g_0 = b/2$, $f_1 = be^{-ic\Delta t/2}/2$, and $g_1 = be^{ic\Delta t/2}/2$.

One can derive continuous functions $f(t) = \lim_{\Delta t \to 0, m\Delta t = t} f_m$ and $g(t) = \lim_{\Delta t \to 0, m\Delta t = t} g_m$ in the following way. The linearlization of Eq. (11) results in

$$\kappa_{m+2} - 2\kappa_{m+1} + \kappa_m - \Delta t \ln x(\kappa_{m+1} - \kappa_m) + (\Delta t)^2 \frac{c^2}{4}\kappa_{m+1}$$

$$- \frac{(\Delta t)^2}{2}(\ln x)^2(\kappa_{m+1} - \kappa_m) + \mathcal{O}[(\Delta t)^3] = 0.$$

Dividing this equation by $(\Delta t)^2$ and taking the limit $\Delta t \to 0$ lead to

$$\partial^2 \kappa(t)/\partial t^2 - \ln x \; \partial \kappa(t)/\partial t + c^2 \kappa(t)/4 = 0,$$

where $\kappa(t) = \lim_{\Delta t \to 0, m\Delta t = t} \kappa_m$. The solution of this differential equation is

$$\kappa(t) = u_\kappa e^{-r_+ t} + v_\kappa e^{-r_- t}$$

with constants u_κ and v_κ ($\kappa = f$ or g), and the complex decoherence factor

$$r_\pm = -\left[\ln x \pm \sqrt{(\ln x)^2 - c^2}\right]/2.$$

The real part of $r_\pm$ is nonnegative because $\ln x = \ln(1 - p)/\tau \leq 0$ and $|\ln x| \geq |\mathrm{Re}\sqrt{(\ln x)^2 - c^2}|$. Here, the square root is positive.

We need to have the conditions $\kappa(0) = b/2$ and $\kappa'(0) = \lim_{\Delta t \to 0}(\kappa_1 - \kappa_0)/\Delta t$ satisfied. The latter condition can be written as $-r_+ u_f - r_- v_f =$

$-ibc/4$ and $-r_+u_g - r_-v_g = ibc/4$. With these conditions, we obtain

$$u_f = \frac{ibc - 2br_-}{4(r_+ - r_-)}, \quad v_f = \frac{-ibc + 2br_+}{4(r_+ - r_-)},$$

$$u_g = \frac{-ibc - 2br_-}{4(r_+ - r_-)}, \quad v_g = \frac{ibc + 2br_+}{4(r_+ - r_-)}.$$

Consequently, we have

$$f(t) = \frac{ibc - 2br_-}{4(r_+ - r_-)}e^{-r_+t} + \frac{-ibc + 2br_+}{4(r_+ - r_-)}e^{-r_-t},$$

$$g(t) = \frac{-ibc - 2br_-}{4(r_+ - r_-)}e^{-r_+t} + \frac{ibc + 2br_+}{4(r_+ - r_-)}e^{-r_-t}.$$

We can now write the reduced density matrix of the principal system at t as

$$\rho^S(t) = \begin{pmatrix} a & \eta(t) \\ \eta(t)^* & 1-a \end{pmatrix},$$

where

$$\eta(t) = b\left(\frac{-r_-}{r_+ - r_-}e^{-r_+t} + \frac{r_+}{r_+ - r_-}e^{-r_-t}\right). \tag{12}$$

This equation describes an exponential decay in the transverse magnetization; the T_2-time relaxation has been achieved.

Let us investigate $r_\pm$ in details in relation to p. One can easily find that the exponential decay is caused by the real part of $e^{-r_\pm t}$. Therefore Re $r_\pm$ are the important factors. We find that Re r_+ increases as $-(\ln x)/2$ when $0 \le p \le 1 - e^{-c\tau}$ and decreases with the convergence to zero when $1 - e^{-c\tau} < p \le 1$. In contrast, Re r_- increases as $-(\ln x)/2$ when $0 \le p \le 1 - e^{-c\tau}$ and starts a rapid growth when $1 - e^{-c\tau} < p \le 1$. These facts are clearly illustrated in Fig. 5. This result suggests that the decoherence factor Re r_+ is small for a large dissipation rate $p > 1 - e^{-c\tau}$. It also suggests that Re e^{-r_-} exhibit a rapidly convergence to zero for $p > 1 - e^{-c\tau}$. Thereby, the dominant term in $\eta(t)$ for a large p is found to be $b\frac{-r_-}{r_+ - r_-}e^{-r_+t}$; this converges to b.

This finding has a physical meaning: the system E is wiped out quickly before affecting coherence information of the system S as p approaches to unity. This effect is a sort of Zeno-like effects and may be called *quantum wipe effect*. It is also regarded as a motional narrowing because a rapid replacement of the localized environment is usually due to a motion of molecules for non-solid samples.

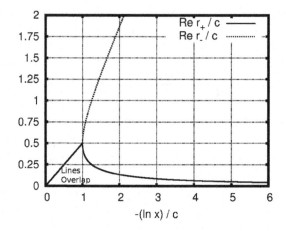

Fig. 5.   Plot of Re $r_\pm/c$ as functions of $-(\ln x)/c$.

The Fourier spectrum of FID is calculated from Eq. (12). It is given by

$$S(\omega) \equiv \mathcal{F}[e^{j\omega_L t}\eta(|t|)]$$

$$= \begin{cases} \frac{2br_+r_-}{r_+-r_-}\left[\frac{-1}{r_+^2+(\omega-\omega_L)^2} + \frac{1}{r_-^2+(\omega-\omega_L)^2}\right] & (r_\pm > 0) \\ \delta(\omega-\omega_L) & (r_+ = 0 \text{ or } r_- = 0) \\ \text{divergent} & (r_\pm = 0) \end{cases} \quad (13)$$

To see the spectrum numerically, let us consider real b and set the parameter values as $\tau = 1.0 \times 10^{-3}$s and $c = 1.0 \times 10^3$Hz. The spectrum for several different p is shown in Fig. 6. The figure clearly depicts the motional narrowing: We have a normal sharp NMR spectrum for p close to zero. There are two peaks because of the spin-spin coupling. The two peaks become rather broad for p in the middle range. They merge to each other and becomes a single peak as p grows further. Finally we have a single sharp peak at the Larmor frequency for p close to one.

It has been shown that the transverse relaxation and the motional narrowing are derived from the single semi-classical model. This has been achieved by a simple spin-spin coupling case. More complicated cases involving spin-boson couplings can be numerically evaluated. The numerical analyses for these cases were reported in Refs.[5–7] Here, as an application, we evaluate the decoherence of nonclassicality in correlations in the model when the principal system has two or three qubits, in the following section.

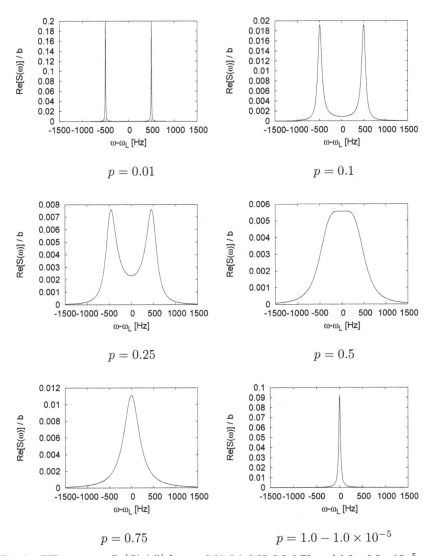

Fig. 6. FID spectrum $\mathrm{Re}[S(\omega)/b]$ for $p = 0.01, 0.1, 0.25, 0.5, 0.75$, and $1.0 - 1.0 \times 10^{-5}$. See the text for the fixed values of the other parameters.

4. Numerical evaluation of dynamical control of decoherence in nonclassical correlation

In this section, we study a decoherence control for a spin-boson linear coupling model among many decoherence models that have been proposed for different physical systems.[11] The spin-boson linear coupling model describes spin energy state transition involving absorption and emission of bosons. This model has been extensively studied in terms of decoherence in single-spin superposition and/or multi-spin entanglement[43] in the light of the recent broad interest in quantum computation.

Suppression of decoherence is indispensable for stable quantum computing and has been studied in several different methodologies. One of them is dynamical decoupling, or bang-bang control[44] (see, e.g., Ref. 45 for recent progress), which has a long history together with the development of nuclear magnetic resonance.[8] The advantage of this method in comparison to the other methods (quantum error correcting codes,[46] decoherence free subspace,[47] etc.) is that it does not require a redundant state space to stabilize the principal system against noise. A series of regular short pulses applied to the principal system cancels noise that has a longer interval time than the applied pulses. This method is often categorized in the same methodology as the Zeno control.[48,49]

Here, we numerically study the bang-bang control of nonclassical correlation and that of entanglement in a sort of spin-boson linear coupling models as we will see in Sec. 4.2.1. The model[5] consists of the principal spin system (spins 1/2) and the environmental bosonic system localized around the principal system. The localized environmental system is probabilistically replaced with a thermal environmental system with a certain dissipation rate. With this model, one can formulate exponential dephasing easily in case of on-resonance spin-boson coupling. We numerically evaluate the effect of coherence conservation in nonclassical correlation that we briefly introduce in Sec. 4.1 under a common bang-bang pulse irradiation. Conservation of entanglement is also evaluated, in addition to that of nonclassical correlation, using a common entanglement measure.

4.1. Nonclassical correlation

There are several different paradigms to discuss nonclassical or quantum correlation of a multipartite system. The most common one is the LOCC paradigm (LOCC is the acronym of local operations and classical communications). This paradigm is related to the separability paradigm.[50] Suppose that there are n participants to make an n-partite quantum state from scratch using LOCC among them. Then, they can create a state represented by a separable density matrix

$$\rho_{\text{sep}} = \sum_j w_j \rho_j^1 \otimes \cdots \otimes \rho_j^n$$

with positive weights w_j ($\sum_j w_j = 1$) and local density matrices ρ_j^k ($k = 1, \ldots, n$). They cannot create a state described by inseparable density matrix, namely a density matrix that cannot be represented by the above form. Entanglement is defined as inseparability.

The separability paradigm has a focus on the state preparation. It does not state that a state after preparation under LOCC is always only classically correlated. To make a mixed state with LOCC, it is needless to say that the order of temporarily prepared states is obliterated so that the process is a mixing process. Thus some separable states in the stage after preparation exhibit nonclassical correlation due to the oblivion of information. Local measurements cannot distinguish some separable states completely not to mention inseparable states. There are many different approaches to characterize such nonclassical correlation: a certain nonlocality about locally nonmeasurable separable states,[34] quantum discord,[51] deficit,[52] etc.

Here we use a measure of nonclassical correlation based on the paradigm proposed by Oppenheim and the Horodecki family.[52] It states that a quantum multipartite system represented by a density matrix having a product eigenbasis (PE)

$$\rho_{\text{PE}} = \sum_{j,k,\ldots,y=1,1,\ldots,1}^{d^1,d^2,\ldots,d^n} c_{j,k,\ldots,y} |e_j^1\rangle\langle e_j^1| \otimes |e_k^2\rangle\langle e_k^2| \otimes \cdots \otimes |e_y^n\rangle\langle e_y^n|$$

with $\{|e_s^m\rangle\}_{s=1}^{d^m}$ an eigenbasis of the density matrix of the mth part (its state space has the dimension of d^m) is regarded as classically correlated. Otherwise the system is regarded as nonclassically correlated. The set of density matrices having no product eigenbasis includes the set of entangled states (the set of entangled states is the complement of the set of separable states) as shown in Fig. 7. Effective QIP requires quantumness to be stable against noise in general. A controversy on quantumness often arises in quantum computing with mixed (and often ensemble) states.[53] A typical example is a pseudo-entangled state

$$\rho_{\text{PS}} = \epsilon|\psi\rangle\langle\psi| + (1 - \epsilon)I/d$$

with $0 < \epsilon \le 1$, $|\psi\rangle$ an entangled pure state, and d the dimension of the Hilbert space. This state is separable for small ϵ[54] but behaves as an entangled state under quantum operations and measurements as long as we neglect the identity matrix. It has no product eigenbasis as long as $\epsilon > 0$; hence it possesses nonclassical correlation. In this sense, NMR quantum computing starting from a pseudo-pure state generated from low-polarized states is still considered a computing using quantumness in correlation.

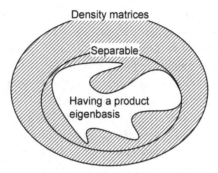

Fig. 7. Set of density matrices having no product eigenbasis (shaded region). Its complementary set (the set of classically-correlated states) is non-convex.

Measures of nonclassical correlation based on Oppenheim and Horodeckis' paradigm have been proposed by their group,[52] Groisman et al.[55] and our group.[56] Among these measures proposed so far, we use the one defined as a minimum discrepancy between an achievable information using local measurements and the information of the total system. This is equivalent to the zero-way quantum deficit proposed by Oppenheim and Horodecki,[52] for the bipartite case. Let us consider a density matrix $\rho^{[1,\dots,n]}$ of an n-partite system. Consider local complete orthonormal bases $\{|e_j^1\rangle\}_j, \dots, \{|e_y^n\rangle\}_y$. Local measurements using observables $M^1 = \sum_j j|e_j^1\rangle\langle e_j^1|, \dots, M^n = \sum_y y|e_y^n\rangle\langle e_y^n|$ are considered. Then, a measure of nonclassical correlation is defined by

$$D(\rho^{1,\dots,n}) = \min_{\text{local bases}} \left(- \sum_{j,\dots,y} p_{j,\dots,y} \log_2 p_{j,\dots,y} \right) - S_{\text{vN}}(\rho^{1,\dots,n})$$

with $p_{j,\dots,y} = \langle e_j^1|\langle e_k^2| \cdots \langle e_y^n|\rho^{1,\dots,n}|e_j^1\rangle|e_k^2\rangle \cdots |e_y^n\rangle$. The value of $D(\rho^{1,\dots,n})$ is zero if $\rho^{1,\dots,n}$ has a (fully) product eigenbasis. In addition, $D(\rho^{1,\dots,n})$ is invariant under local unitary operations as is clear from its definition. We use a numerical random search method to compute the value of D for a given density matrix.[56]

4.2. Bang-bang control in a spin-boson model

4.2.1. On-resonance spin-boson linear coupling with environmental dissipation

There is a longstanding study of suppressing decoherence in quantumness. Let us focus on the bang-bang control scheme.[44] The decoherence model we investigate is a sort of spin-boson linear coupling models (see, e.g., Ref. 11 for conventional models). We consider a spin system S (consisting of n spins 1/2) coupled with an on-resonance bosonic system B that is replaced with a thermal bosonic system with dissipation probability p for a time interval τ. This model is illustrated as Fig. 4 with the bosonic environmental system indicated by symbol B instead of E in the figure.

The system Hamiltonian is set to $H = H_S + H_B + H_c$ with $H_S = \sum_{j=1}^{n} f_j S_j + \sum_{j,k,j<k} A_{jk} S_j S_k$ (spin system Hamiltonian), $H_B = \sum_{j=1}^{n} f_j a_j^\dagger a_j$ (Hamiltonian of bosonic modes), and $H_c = \sum_{j=1}^{n} c_j S_j (a_j^\dagger + a_j)$. Here, f_j is the precession frequency of the jth spin ($f_j \neq f_k$ for $j \neq k$), A_{jk} is a spin-spin coupling constant, c_j is a spin-boson on-resonance coupling constant.

With a small time interval Δt, the system evolution is computed directly from the equation

$$\rho^{SB}(\tilde{t} + \Delta t) = e^{-iH\Delta t}\left[x^{\Delta t}\rho^{SB}(\tilde{t}) + (1 - x^{\Delta t})\mathrm{Tr}_B\rho^{SB}(\tilde{t}) \otimes \sigma\right]e^{iH\Delta t}, \quad (14)$$

where $x = (1 - p)^{1/\tau}$ and $\tilde{t}$ denotes a certain time step and $\sigma = e^{-\beta H_B}/Z_B$ with Z_B the partition function.

4.3. Bang-bang control

We consider a bang-bang pulse irradiation with pulse width L and duty ratio R [i.e., one X_π pulse is irradiated with the duration of RL followed by the absence time interval $(1 - \text{R})L$]. The bang-bang pulses for individual spins are all synchronized. In simulation, this effect is computed by adding the Hamiltonian $H_X = \sigma_x \pi/(2\text{R}L)$ of the pulse to H for every duration of the irradiation.

We are going to numerically evaluate the effect of bang-bang pulses on time evolutions of a measure of non-classical correlation and a measure of entanglement.

4.4. Numerical results

For the principal system S, let us consider a solid sample including molecules with an electron spin (labeled as 1) and a nuclear spin (labeled as 2) coupled with the hyperfine coupling constant A_{12} (hence S=12). An ENDOR quantum computing[57,58] in the Q-band (namely, the system with the electron-spin precession frequency approximately 34GHz) is tacitly assumed in the setup. We numerically evaluate the effect of bang-bang control to suppress decoherence in our model as we have mentioned in previous subsections. The computer system we use has two Intel Xeon E5345 processors (2.33GHz four cores for each) and 16GB physical memory. For this simulation, we set the following parameter values: $f_1 = 3.40 \times 10^{10}$[Hz], $f_2 = 4.87 \times 10^7$[Hz], $A_{12} = 1.00 \times 10^7$[Hz], $c_1 = c_2 = 1.00 \times 10^3$[Hz], temperature 1.00mK, $\tau = 1.00 \times 10^{-6}$[s], $L = 1.00 \times 10^{-7}$[s], $\Delta t = 5.00 \times 10^{-8}$[s] and $p = 0.10$. We apply an entangling operation consisting of the Hadamard gate acting on the electron spin followed by the CNOT gate (with the control bit in the electron spin) to the spins originally in the thermal state $\rho^{SB}(0) = e^{-\beta H}/Z$ (Z is the partition function). The state after this operation is the initial state in the simulation of decoherence and its suppression. Time evolutions of D and negativity[59] (one of entanglement measures. See Footnote [*1] for its definition.) with/without bang-bang control are shown in Figs. 8 and 9. Computation to make a single plot of D involved 3.80×10^4 randomly generated product complete orthonormal bases to take the minimum value. Suppression of decoherence is significant for the duty ratio $\textsc{r} = 1.00 \times 10^{-4}$. In the dynamical control of negativity, an anti-Zeno effect[60] (namely, a growth of decoherence owing to an insufficient dynamical control) is observed for $\textsc{r} = 1.00 \times 10^{-2}$ as shown in Fig. 9. The nonclassical correlation D is, in contrast, conserved well for the same value of $\textsc{r}$ as shown in Fig. 8.

Let us change the condition slightly to perform another simulation. We consider a similar model with system S consisting of an electron spin (labeled as 1) and two nuclear spins (labeled as 2 and 3). The electron spin is coupled with the nuclear spins with the hyperfine coupling constants $A_{12} = A_{13} = 1.00 \times 10^7$[Hz]; A_{23} is set to zero. We use $L = 2.00 \times 10^{-8}$[s] and $\Delta t = 1.00 \times 10^{-8}$[s]. We keep the other parameters unchanged except for the precession frequency $f_3 = f_2 + 1.00 \times 10^2$[Hz] and the spin-boson coupling constant $c_3 = 1.00 \times 10^3$[Hz] assigned to the newly-added nuclear spin. We start with the thermal state and apply the Hadamard gate to the electron spin followed by two CNOT gates with the control bit in the electron spin and the target bit in each nuclear spin. Time evolution starting from this stage was computed with/without the bang-bang control. The same measure of nonclassical correlation is used while the entanglement measure is changed to the minimum negativity over all bipartite splittings.

[*1]The negativity of a system for a bipartite splitting between subsystems A and B is $\mathcal{N}(\rho^{A,B}) = [\|(I \otimes \Lambda_{\text{T}})\rho^{A,B}\| - 1]/2$ (here the map Λ_{T} is the transposition map acting on B).

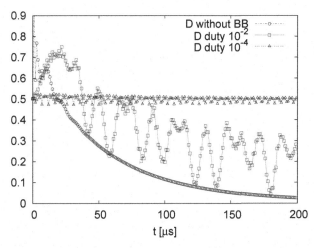

Fig. 8. Time evolutions of D with/without bang-bang control for the model involving a couple of spins $1/2$ in the principal system. Duty ratios R $= 1.00 \times 10^{-2}$ and R $= 1.00 \times 10^{-4}$ were tried.

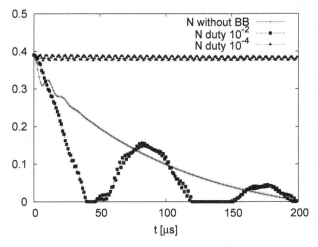

Fig. 9. Time evolutions of negativity (N) with/without bang-bang control for the model involving a couple of spins $1/2$ in the principal system. Duty ratios R $= 1.00 \times 10^{-2}$ and R $= 1.00 \times 10^{-4}$ were tried.

Figures 10 and 11 shows faster decoherence than the previous case. As we see in Fig. 11, the duty ratio 1.00×10^{-2} of the bang-bang pulse seems insufficient and results in an acceleration of decoherence in negativity. This is again an anti-Zeno effect. In contrast, the same duty ratio provides a better, albeit not effective, coherence conservation of nonclassical correlation D as shown in Fig. 10. As depicted in these figures, it is found that coherence

174

conservation is not effective even with the smaller duty ratio 1.00×10^{-4}. The oscillation is rather suppressed while the decay of the amplitude of the minimum negativety and that of D are enhanced rather than suppressed. In particular for D, the smaller duty ratio seemingly results in a worse effect than the larger one[*2].

4.5. *Discussions on the simulation results*

We chose a bulk-ensemble spin system under low temperature for the principal system in the numerical evaluation. We have investigated the effect of bang-bang control for suppressing decoherence numerically for Q-band ENDOR-like systems. It is expected that nonclassical correlation sustains longer than entanglement under noise since the set of nonclassically correlated states is a super set of the set of inseparable states. The difference in the decay time of nonclassical correlation and that of entanglement is significant in the case where the duty ratio of a regular pulse of bang-bang control is not small enough to achieve effective suppression of decoherence according to the simulation result, shown in Figs. 8 and 9. The principal system consisting of two spins $1/2$ is stabilized when the duty ratio is 1.00×10^{-4} with the pulse interval 1.00×10^{-7}, namely when the pulse duration is 0.01ns. This is, however, not realistic with the currently available technology. Nonclassical correlation sustains long when the duty ratio is 1.00×10^{-2}, namely when the pulse duration is 1.00ns. This is within a realistic range of pulse duration available for ENDOR machines in forthcoming years.

Suppressing decoherence for the system with three spins $1/2$ is a severe challenge as we have seen in Figs. 10 and 11. It is not possible to achieve suppression even by applying very short pulses with the interval 2.00×10^{-8} and the duty ratio 1.00×10^{-4}, namely, with the pulse duration 2.00ps. This is far beyond the realistic range available for the current magnetic resonance technology. Thus it suggests that the bang-bang control to suppress decoherence in nonclassical correlation due to spin-boson linear coupling is effective only for the system consisting of no more than two spins. Other techniques like error correcting codes should be tried to accomplish the task for systems with more than two spins.

To investigate decoherence in bulk ensemble spin systems faithfully,

---

[*2]We have so far no plausible explanation of this phenomenon. There is a possibility that nonclassical correlation is partly generated by a bang-bang control with a certain duty ratio together with the spin-boson coupling.

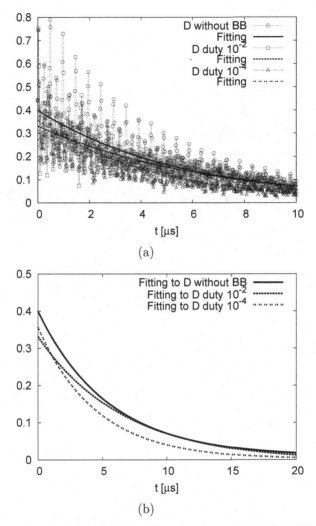

Fig. 10. Time evolutions of D with/without bang-bang control for the model involving three spins $1/2$ in the principal system. Duty ratios R $= 1.00 \times 10^{-2}$ and R $= 1.00 \times 10^{-4}$ were tried. (a) Data points with curves obtained by least square fitting ($0 \leq t \leq 10$ [μs]); (b) The curves only ($0 \leq t \leq 20$ [μs]). The exponential decay $Ae^{-Bt} + C$ is assumed for the fitting, with the fitting parameters $A, B, C \in \mathbf{R}$.

spin-spin coupling should be considered in addition to spin-boson coupling. In a solid sample, inter-molecule spin-spin coupling is non-negligible in addition to intra-molecule spin-spin coupling. Analyses of such systems using

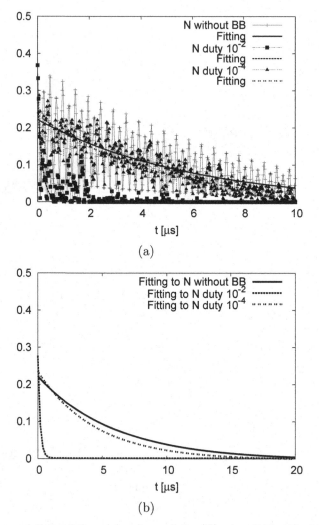

Fig. 11.   Time evolutions of the minimum negativity (N) over all bipartite splittings with/without bang-bang control for the model involving three spins 1/2 in the principal system. Duty ratios R $= 1.00 \times 10^{-2}$ and R $= 1.00 \times 10^{-4}$ were tried. (a) Data points with curves obtained by least square fitting ($0 \leq t \leq 10$ [μs]); (b) The curves only ($0 \leq t \leq 20$ [μs]). The exponential decay $Ae^{-Bt} + C$ is assumed for the fitting, with the fitting parameters $A, B, C \in \mathbf{R}$.

differential equations pertaining to population transitions among energy eigenstates is very common.[40] It is a future task to extend the formulation of Eq. (9) so that a time evolution of a spin-system density matrix under spin-spin coupling in addition to spin-boson coupling is numerically computed.

5. Concluding remarks

It has been shown that a transverse relaxation and a motional narrowing of a spin 1/2 due to a spin-spin coupling are both elucidated analytically in the model illustrated in Fig. 4 that is governed by the recursive equation (9). The exponential decay of the transverse magnetic moment is achieved in the analytical result Eq. (12); the FID spectrum obtained as the analytical result, depicted in Fig. 6, clearly exhibits the process of motional narrowing. This suggests that a larger dissipation rate in the environment results in a better conservation of the coherence of the system in our model; the environment is wiped out before it leaves much influence to the system. Thus we may call this effect a quantum wipe effect. The recursive equation uses the factor $1 - (1 - p)^{\Delta t/\tau}$ for the dissipation probability for small time duration Δt. This is the essential difference between our custom and the conventional custom in which the dissipation probability is written as $p\Delta t/\tau$. It is clear that ours is more appropriate in view of the basic probability theory. We expect that our custom will also apply for the theory of quantum error propagation to achieve more accurate error estimations.

We have investigated the decay of quantumness—entanglement and nonclassical correlation— in a Q-band ENDOR-like model using the recursive equation in numerical simulations. Both an effective suppression and an anti-Zeno effect have been observed. Nonclassical correlation has been found to be more robust than entanglement as was expected. The numerical simulations were possible in view of computational costs as we consider only on-resonance coupling between spins and bosonic modes. Although it is computationally difficult to involve off-resonance couplings in general, it might be possible to consider off-resonance coupling with a sharp peak at the resonance frequency. In addition, one may also add environmental spins. These are future tasks to achieve more realistic and reliable simulation results.

References

1. N. Bloembergen, E. M. Purcell and R.V. Pound: Phys. Rev. **73** (1948) 679.
2. M. Ban: J. Mod. Opt. **45** (1998) 2315.
3. L. Viola and S. Lloyd: Phys. Rev. A **58** (1998) 2733.
4. A. SaiToh, R. Rahimi and M. Nakahara: LANL arXiv: 0709.0562 (quant-ph).

178

5. R. Rahimi, A. SaiToh and M Nakahara: J. Phys. Soc. Japan **76** (2007) 114007.
6. R. Rahimi, A. SaiToh and M Nakahara: Int. J. Quant. Inf. **6**, Supp. 1 (2008) 779.
7. R. Rahimi, A. SaiToh and M Nakahara: Iran. Phys. J. **2-3** (2008) 8.
8. R. R. Ernst, G. Bodenhausen and A. Wokaun: *Principles of Nuclear Magnetic Resonance in One And Two Dimensions* (Oxford University Press, Oxford, 1987).
9. P. W. Anderson and P. R. Weiss: Rev. Mod. Phys. **25** (1953) 269.
10. R. Kubo and K. Tomita: J. Phys. Soc. Jpn. **9** (1954) 888.
11. W. H. Louisell: *Quantum Statistical Properties of Radiation* (Wiley, New York, 1973).
12. H.-P. Breuer and F. Petruccione, *The Theory of Open Quantum Systems* (Oxford University Press, Oxford, 2002).
13. J. Gruska: *Quantum Computing* (McGraw-Hill, London, 1999).
14. M. A. Nielsen and I. L. Chuang: *Quantum Computation and Quantum Information* (Cambridge University Press, Cambridge, 2000).
15. C. Bennett and G. Brassard: in *Proceedings of the IEEE Int. Conf. Comput., Sys. & Sig. Proc.*, Bangalore, India, 9-12 Dec. 1984 (IEEE, 1984) 175.
16. D. R. Simon: in *Proceedings of the 35th IEEE FOCS* (1994) 116; SIAM J. Comput. **26** (1997) 1474.
17. P. W. Shore: in *Proceedings of the 35th IEEE FOCS* (1994) 124; SIAM J. Comput. **26** (1997) 1484.
18. C. H. Bennett, G. Brassard, C. Crépeau, R. Jozsa, A. Peres and W. K. Wootters: Phys. Rev. Lett. **70** (1993) 1895.
19. J. Zhou, G. Hou, S. Wu and Y. Zhang: LANL arXiv: quant-ph/0006030.
20. N. B. An: Phys. Rev. A **68** (2003) 022321.
21. C.-P. Yang, S.-I. Chu and S. Han: Phys. Rev. A **70** (2004) 022329.
22. F. G. Deng, C.-Y. Li, Y.-S. Li, H.-Y. Zhou and Y. Wang: Phys. Rev. A **72** (2005) 022338.
23. Z.-X. Man, Y.-J. Xia and N. B. An: J. Phys. B: At. Mol. Opt. Phys. **40** (2007) 1767.
24. A. SaiToh, R. Rahimi and M. Nakahara: Phys. Rev. A **79** (2009) 062313.
25. B. M. Terhal, D. P. DiVincenzo and D. W. Leung: Phys. Rev. Lett. **86** (2001) 5807.
26. D. P. DiVincenzo, D. W. Leung and B. M. Terhal: IEEE Trans. Inf. Theory **48** (2002) 580.
27. D. P. DiVincenzo, P. Hayden and B. M. Terhal: Found. Phys. **33** (2003) 1629.
28. R. Cleve, D. Gottesman and H.-K. Lo: Phys. Rev. Lett. **83** (1999) 648.
29. D. Markham and B. C. Sanders: Phys. Rev. A **78** (2008) 042309.
30. G. Vidal: Phys. Rev. Lett. **91** (2003) 147902.
31. E. Knill and R. Laflamme: Phys. Rev. Lett. **81** (1998) 5672.
32. A. Datta, A. Shaji and C. M. Caves: Phys. Rev. Lett. **100** (2008) 050502.
33. H. Ollivier and W. H. Zurek: Phys. Rev. Lett. **88** (2001) 017901.
34. C. H. Bennett, D. P. DiVincenzo, C. A. Fuchs, T. Mor, E. Rains, P. W. Shor, J. A. Smolin and W. K. Wootters: Phys. Rev. A **59** (1999) 1070.
35. J. Oppenheim, M. Horodecki, P. Horodecki and R. Horodecki: Phys. Rev.

Lett **89** (2002) 180402.

36. B. Groisman, S. Popescu and A. Winter: Phys. Rev. A **72** (2005) 032317.
37. T. Yu and J. H. Eberly: Phys. Rev. Lett. **97** (2006) 140403.
38. A. Abragam: *Principles of Nuclear Magnetism* (Oxford University Press, New York, 1961).
39. M. H. Levitt: *Spin Dynamics* (Wiley, Chichester, 2001).
40. R. Kitamaru: *Basics and Principles of Nuclear Magnetic Resonance* (Kyoritsu, Tokyo, 1987, in Japanese).
41. M. Matsuo: Prog. Theor. Phys. **84** (1990) 269.
42. N. D. Dang and F. Sakata: Phys. Rev. C **57** (1998) 3032.
43. D. Mozyrsky and V. Privman: J. Stat. Phys. **91** (1998) 787; D. Tolkunov, V. Privman and P. K. Aravind: Phys. Rev. A **71** (2005) 060308(R).
44. M. Ban: J. Mod. Opt. **45** (1998) 2315; L. Viola and S. Lloyd: Phys. Rev. A **58** (1998) 2733.
45. K. Shiokawa and B. L. Hu: Quant. Inf. Processing **6** (2007) 55.
46. P. W. Shor: Phys. Rev. A **52** (1995) R2493; D. Gottesman: Phys. Rev. A **54** (1996) 1862; R. Laflamme, C. Miquel, J. P. Paz and W. H. Zurek: Phys. Rev. Lett. **77** (1996) 198.
47. P. Zanardi and M. Rasetti: Phys. Rev. Lett. **79** (1997) 3306; L.-M. Duan and G.-C. Guo: Phys. Rev. Lett. **79** (1997) 1953; D. A. Lidar, I. L. Chuang and K. B. Whaley: Phys. Rev. Lett. **81** (1998) 2594.
48. B. Misra and E. C. G. Sudarshan: J. Math. Phys. **18** (1977) 756.
49. P. Facchi and S. Pascazio: Phys. Rev. Lett. **89** (2002) 080401.
50. A. Peres: Phys. Rev. Lett. **77** (1996) 1413.
51. H. Ollivier and W. H. Zurek: PhysRev. Lett. **88** (2001) 017901.
52. J. Oppenheim, M. Horodecki, P. Horodecki and R. Horodecki: Phys. Rev. Lett. **89** (2002) 180402; M. Horodecki, P. Horodecki, R. Horodecki, J. Oppenheim, A. Sen(De), U. Sen and B. Synak-Radtke: Phys. Rev. A **71** (2005) 062307.
53. E.g., discussions after the talk $WE - 2$ of A. G. White *et al.* in the 9th Int. Conf. Quant. Commu., Measur. & Comput., Calgary, Canada, 19-24 August 2008.
54. S. L. Braunstein, C. M. Caves, R. Jozsa, N. Linden, S. Popescu and R. Schack: Phys. Rev. Lett. **83** (1999) 1054.
55. B. Groisman, D. Kenigsberg and T. Mor: e-print quant-ph/0703103.
56. A. SaiToh, R. Rahimi and M. Nakahara: Phys. Rev. A **77** (2008) 052101.
57. M. Mehring, J. Mende and W. Scherer: Phys. Rev. Lett. **90** (2003) 153001.
58. R. Rahimi: PhD Thesis, Osaka University, Osaka, (2006) e-print quant-ph/0609063.
59. K. Życzkowski, P. Horodecki, A. Sanpera and M. Lewenstein: Phys. Rev. A **58** (1998) 883; G. Vidal and R. F. Werner: Phys. Rev. A **65** (2002) 032314.
60. A. G. Kofman and G. Kurizki: Nature (London) **405** (2000) 546.

HOLONOMIC QUANTUM GATES
USING
ISOSPECTRAL DEFORMATION OF ISING MODEL

MASAMITSU BANDO[1], YUKIHIRO OTA[2]*, YASUSHI KONDO[2,3]

and

MIKIO NAKAHARA[2,3]

[1]*Department of Physics, Graduate School of Science, Osaka University,*
1-1 Machikaneyama, Toyonaka, Osaka 560-0043, Japan
[2]*Research Center for Quantum Computing,*
Interdisciplinary Graduate School of Science and Engineering,
Kinki University, 3-4-1 Kowakae, Higashi-Osaka, 577-8502, Japan
[3]*Department of Physics, Kinki University,*
3-4-1 Kowakae, Higashi-Osaka, 577-8502, Japan

We outline the theory of holonomic quantum computing and implementation of
a holonomic quantum gate. Then we analyze an isospectral deformation of the
Ising-type model Hamiltonian as an example of holonomic quantum computing
and construct a few holonomic quantum gates using this system.

Keywords: Quantum Computing; Holonomy; Ising Model.

1. Holonomic Quantum Computing

Let us consider a family of Hamiltonians whose parameter manifold is M
and take a closed loop $\mathcal{C}$ in M. The parameters change along the loop $\mathcal{C}$ as
a function of time. We write the resulting Hamiltonian as $H(t)$ instead of
$H(\mathcal{C}(t))$ to simplify the notation. Suppose that the Hamiltonian $H(0)$ has
g-fold degenerate ground state. We assume that the initial state is expressed
as a superposition of the ground state eigenvectors. When the parameters
are changed adiabatically along this loop, the resulting quantum state at
the end is different from the initial state by an action of a unitary matrix
$\Gamma[\mathcal{C}] \in U(g)$, where we assume that no level-crossing involving the ground
state takes place. This matrix $\Gamma[\mathcal{C}]$ is called holonomy associated with an

*Present address: CCSE, Japan Atomic Energy Agency, 6-9-3 Higashi-Ueno, Tokyo 110-
0015, Japan and CREST(JST), 4-1-8 Honcho, Kawaguchi, Saitama, 332-0012, Japan

182

anti-Hermitian connection, which defines the parallel transport of a vector in the relevant fibre bundle.[1-3] The holonomic quantum computing makes use of such a holonomy to implement a quantum gate.[2-4]

2. Holonomy

Suppose that the quantum system is in the ground state of the Hamiltonian $H(0)$ at $t = 0$. It is assumed that the ground state has the energy E_0 and g-fold degenerate and is separated from the other energy eigenstates by a finite gap. Then the adiabatic time evolution operator associated with the time-dependent Hamiltonian $H(t)$ takes the form

$$\mathcal{T}\left[\exp\left\{-iT\int_0^1 H(\tau)d\tau\right\}\right]P_0 \simeq e^{-iT\int_0^1 E_0(\tau)d\tau}\Gamma P_0,$$

where T is the time which takes to go round the loop $\mathcal{C}$ in M, $\mathcal{T}$ is the time-ordering operator, and P_0 is the projection operator onto the ground state subspace. The matrix $\Gamma \in U(g)$ is called the holonomy associated with the loop.

Now, we consider the isospectral deformation

$$H(\tau) = e^{X\tau}H_0 e^{-X\tau} \qquad (0 \leq \tau \leq 1),$$

where X is an anti-Hermitian operator. It is necessary for implementing a holonomic quantum gate to impose the following two conditions. (i) The Hamiltonian must return to the initial one at $\tau = 1$:

$$e^X = I,$$

where I is the identity operator. (ii) No undesired transitions into the non-coding space should take place.

Using this deformation, we find that the matrix elements of the anti-Hermitian connection A_{ij} are independent of time;

$$A_{ij}(t) = \langle E_0, i| X |E_0, j\rangle = A_{ij}.$$

Now we obtain the holonomy:

$$\Gamma = \mathcal{T}\left[\exp\left\{-\int_0^1 A(\tau)d\tau\right\}\right] = e^{-A}.$$

This equation shows that the holonomy is given by the anti-Hermitian operator A.

3. One-Qubit Quantum Gate

We introduce a Hamiltonian of two-spin system with the Ising type inter-action.[5]

$$H_0 = -\omega \sigma_z \otimes I - \omega I \otimes \sigma_z + J\sigma_z \otimes \sigma_z.$$

Now, we take four basis vectors

$$|T_+\rangle = |++\rangle, \ |T_0\rangle = \frac{1}{\sqrt{2}}(|+-\rangle + |-+\rangle), \ |T_-\rangle = |--\rangle,$$

$$|S_0\rangle = \frac{1}{\sqrt{2}}(|+-\rangle - |-+\rangle).$$

Here the up-spin state is denoted as $|+\rangle$, while the down-spin state as $|-\rangle$. Evaluating the eigenenergies of these basis vectors and taking $\omega = J$, we obtain a three-dimensional ground eigen-subspace spanned by $\{|T_+\rangle, |T_0\rangle, |S_0\rangle\}$ and a single excited state $|T_-\rangle$.

Next, we take the anti-Hermitian operator X as

$$X = i\boldsymbol{n} \cdot \nu (\boldsymbol{\sigma}_1 + \boldsymbol{\sigma}_2),$$

where

$$\boldsymbol{\sigma}_1 = \boldsymbol{\sigma} \otimes I, \quad \boldsymbol{\sigma}_2 = I \otimes \boldsymbol{\sigma}, \quad \boldsymbol{\sigma} = (\sigma_x, \sigma_y, \sigma_z),$$

$\boldsymbol{n}$ is a unit vector in $\mathbb{R}^3$ and ν is a real positive constant. Under this choice of X, the transitions between $|S_0\rangle$ and other states are forbidden and hence we may employ $|T_+\rangle$ and $|T_0\rangle$ as the qubit basis vectors. The anti-Hermitian connection matrix A is then expressed as

$$A = \begin{pmatrix} 2i\nu n_z & \sqrt{2}i\nu(n_x - in_y) & 0 \\ \sqrt{2}i\nu(n_x + in_y) & 0 & 0 \\ 0 & 0 & 0 \end{pmatrix}$$

It is necessary to take $\nu = \kappa\pi$ ($\kappa \in \mathbb{N}$) so that $H((\tau)) = H(0)$. Finally, we obtain the holonomy as

$$\Gamma = e^{-i\kappa\pi n_z} e^{-i\kappa\pi\sqrt{2-n_z^2}\boldsymbol{m}\cdot\boldsymbol{\sigma}}$$

$$= \begin{pmatrix} \cos\theta - im_z\sin\theta & -(im_x + m_y)\sin\theta \\ (-im_x + m_y)\sin\theta & \cos\theta - im_z\sin\theta \end{pmatrix},$$

where

$$\theta = \kappa\pi\sqrt{2 - n_z^2}, \quad \boldsymbol{m} = \left(\frac{\sqrt{2}n_x}{\sqrt{2 - n_z^2}}, \frac{\sqrt{2}n_y}{\sqrt{2 - n_z^2}}, \frac{n_z}{\sqrt{2 - n_z^2}} \right).$$

184

4. Construction of Quantum Gates

Let us take $n_1 = (1, \ 0, \ 0)$ and $\kappa = 1$ first. Then we obtain the gate

$$\Gamma_1 = \begin{pmatrix} \cos(\sqrt{2}\pi) & -i\sin(\sqrt{2}\pi) \\ -i\sin(\sqrt{2}\pi) & \cos(\sqrt{2}\pi) \end{pmatrix}.$$

Next, we take $n_2 = (0, \ 1, \ 0)$ and $\kappa = 1$ to obtain

$$\Gamma_2 = \begin{pmatrix} \cos(\sqrt{2}\pi) & -\sin(\sqrt{2}\pi) \\ \sin(\sqrt{2}\pi) & \cos(\sqrt{2}\pi) \end{pmatrix}.$$

These gates do not commute with each other. Repeated applications of these gates produce a dense subset of matrices in the SU(2) group. This shows that any one-qubit gate may be implemented as a holonomic gate with an arbitrary precision.

The Hadamard gate is implemented with a good accuracy by putting $m = \frac{1}{\sqrt{2}}(1, \ 0, \ 1)$, and select proper κ values, $\kappa = 3, 10$ and 16 for example.

5. Summary and Discussion

We have demonstrated that an arbitrary SU(2) gate may be implemented with an arbitrary precision as a holonomic quantum gate using isospectral deformation of the Ising-like Hamiltonian. A similar construction applies to a two-qubit gate.[5] It is desirable to implement the gates introduced in this work with a real physical system, such as an NMR quantum computer.

References

1. M. Nakahara: *Geometry, Topology and Physics*, 2nd. ed. (IOP Publishing, Bristol and Philadelphia, 2003).
2. K. Fujii: J. Math. Phys. **41** (2000) 4406.
3. S. Tanimura, M. Nakahara and D. Hayashi: J. Math, Phys. **46** (2005) 022101.
4. P. Zanardi and M. Rasetti: Phys. Lett. A **264**, (1999) 94.
5. Y. Ota, M. Bando, Y. Kondo and M. Nakahara: Phys. Rev. A **78** (2008) 052315.

INDEX